CAREERS IN

TELEVISION
AND RADIO

CAREERS IN

TELEVISION AND RADIO

Michael Selby

seventh edition

**KOGAN
PAGE**

WITHDRAWN

First published in 1986, entitled *Careers in Radio and Television* by Susan Crimp
Second edition 1988, entitled *Careers in Television and Radio* by Julia Allen
Third edition 1990
Fourth edition 1993
Fifth edition 1994, by Michael Selby
Sixth edition 1996
Seventh edition 1997

Kogan Page Limited
120 Pentonville Road
London N1 9JN

© Kogan Page 1986, 1988, 1990, 1993, 1994, 1996, 1997

British Library Cataloguing in Publication Data

A CIP record for this book is available from the British Library.

ISBN 0 7494 2421 4

Typeset by Kogan Page
Printed and bound in Great Britain by Clays Ltd, St Ives plc

Contents

Acknowledgements

I would like to thank everyone who contributed to this update in any way, particularly those who responded in detail to my letters and faxes, suggested improvements to the book or who updated the text from the previous edition. I would also like to thank the industry 'names' who took the time to offer a few words of advice for the 'Tips from the Top' section, and especially the subjects of the case studies. I am most grateful to all who contributed. I must also acknowledge that I have made extensive use of the *Skillset Careers Handbook*, published by Skillset, and the *Official ITV Careers Handbook*, published by Headway/Hodder & Stoughton.

Finally, a very special thank you to Shiona Llewellyn for her input, and to John Bennett, Joanna Hughes, Owen McFadden and Fiona McGaughey for their unfailing help and support.

Michael Selby

Introduction

Is this the Job for You?

☐ *Why do you want to work in the media?*
If it's just because you imagine a glamorous, exciting, well-paid lifestyle... forget it!

☐ *What do you want to do?*
Think carefully – it helps if you have at least a fair idea of the area(s) in which you want to work.

☐ *What are your strengths and weaknesses?*
You have to be honest about your own abilities. Realistically, only the most able, creative, dedicated and enthusiastic candidates are likely to succeed.

☐ *What skills do you have – and which ones would be useful for you to acquire?*
Core qualities are self-motivation, flexibility and communication skills.

☐ *Are you prepared to work on a freelance basis with the potential insecurities this brings?*

☐ *Are you prepared to travel to find work?*

☐ *Are you prepared to work long hours for little or no money to gain experience?*

☐ *Can you work as part of a team?*
Successful programmes depend on the effective cooperation of everyone involved.

❏ *Can you keep a cool head under pressure?*
Deadlines must be adhered to if a programme is to be completed on time and within budget – and this can lead to enormous pressure.

❏ *Are you prepared to stick at it?*
Success is (almost) never instantaneous. Don't expect to land that dream job in the first few months, or even years, of embarking on a media career.

What are your strengths and weaknesses?

Anybody can say they want to work in television or radio; the problem is that there are thousands of people with the same vague 'media' ambitions, many of whom would happily sell their grandparents to secure that first job.

Although it may seem that new channels are springing up everywhere, new jobs are not, unfortunately, appearing at a similar rate. Certainly, the entire television industry is on the verge of a new era with the launch of digital broadcasting. However, many of the new channels on offer will be restricted by small budgets, while others will rely heavily on archive material.

Technological advances have also had a major impact behind the scenes. New developments have resulted in job cuts in most fields, with the result that the workforce is becoming increasingly multi-skilled.

Meanwhile, the cable and satellite sectors continue to thrive, and there has also been good news for independent producers, with the launch of Channel 5. The new station – the last terrestrial network before the arrival of digital – commissions the bulk of its output from independent production companies.

Major challenges continue to confront the more established broadcasters: the BBC has been reorganised and restructured, while ITV, once a network of 15 individual companies, is now dominated by three powerful media groups. This has resulted in a rationalisation of resources with the inevitable redundancies, and further numbers entering the freelance market.

The national radio networks – both commercial and BBC – have been struggling to hold on to their existing audiences as more local stations take to the airwaves, while Digital Audio Broadcasting, which is due to launch before the end of the century, will lead to a further expansion in the number of services.

The broadcast medium is a tough business and the competition is fierce (a phrase you will encounter more than once in this book!) The opportunities do still exist – the trick is knowing where to look for them and how to make the most of them. The best advice is to be realistic about your own strengths and weaknesses, keep informed and, above all, persevere.

You will find complementary information in other books in this series, *Careers in Film and Video* and *Careers in Journalism*.

2 The employers

The BBC

The BBC is by far the biggest employer among the broadcasting organisations. It is concerned with all aspects of broadcasting, including news gathering, programme making and technical innovation. Domestically, the BBC operates two national television channels, BBC1 and BBC2, and five national radio stations: Radios 1, 2, 3, 4 and 5 Live. Further services are scheduled to launch over the next couple of years, as digital broadcasting develops (see page 60).

The Corporation has three 'National Regions' – Scotland, Wales and Northern Ireland – and nine English regional centres, collectively known as the 'English Regions': Birmingham, Bristol, Leeds, Manchester, Newcastle upon Tyne, Norwich, Nottingham, Plymouth and Southampton. There are also production bases in Borehamwood (the Elstree Centre) and Milton Keynes (the Open University Production Centre). The BBC operates 39 local radio stations across England, plus regional radio services in each of the national regions.

Information about career opportunities can be obtained from BBC Corporate Recruitment Services (either contact the London address on page 71 or the personnel department of your local BBC), although there is intense competition for jobs and traineeships with the Corporation.

Extending Choice

The BBC's right to broadcast and to collect the licence fee is granted by Parliament through a Royal Charter. This is reviewed – and has so far been renewed – every ten years and, in effect, means that the Corporation is continually forced to justify its existence.

In preparation for the 1996 Charter renewal, a number of initiatives were instigated under the banner 'Extending Choice' which were designed, at least in theory, to transform the BBC into a more streamlined, competitive business operation. These measures included the introduction of an internal market within the Corporation – which means that departments must now sell their services to each other, as they would to outside clients. Producers, for example, are no longer obliged to use BBC resources if a more cost-effective facility is available outside the Corporation.

As part of this overall efficiency drive, the BBC – like many other organisations – aims to slim down to a core number of permanent staff, recruiting further numbers on a fixed-term contract basis as work demands. This policy has resulted in thousands of job losses over the past few years, although the situation is beginning to stabilise.

The structure of the BBC

BBC Broadcast is responsible for commissioning and scheduling the Corporation's entire output, with the exception of news. Its remit covers: BBC1 and BBC2; Radios 1, 2, 3 and 4 (but not 5 Live); Education; all local and regional services across the UK; and the new digital services, both commercial and licence-fee funded.

BBC Production, the Corporation's programme-making division, supplies programmes and services to all BBC radio and television outlets at home and abroad. Production employs some 4,000 people, most of whom are based in the regions.

BBC News is the world's largest news-gathering organisation. Domestic news is produced by four specialist units – Politics,

Economics, Business & Social Affairs and Foreign Affairs – while foreign reports are gathered from the Corporation's 26 bureaux around the globe. BBC News provides coverage for all BBC services, including BBC1 and 2, Radios 1, 2, 3, 4 and 5 Live, BBC World (television) and BBC World Service (radio).

BBC Resources incorporates departments and services such as outside broadcast facilities, camera crews, studios, post-production, design and archives. The BBC ultimately aims to 'spin-off' Resources into a wholly owned subsidiary company.

BBC Worldwide

BBC Worldwide is the international and commercial arm of the BBC. It consists of two main business areas: television (including cable and satellite channels, and the international distribution of BBC programmes); and publishing (including books, magazines, videos, audio tapes and new media).

One non-commercial division of Worldwide is the *BBC World Service*. Funded by a Parliamentary grant, the World Service operates under the BBC's Royal Charter and is managerially independent of the government.

The Independent Television Network (ITV)

ITV – sometimes referred to as Channel 3 – is a network of individual companies, each broadcasting to a specific area or 'region' of the country. In practice, these regional boundaries are fairly arbitrary (most have been altered to some degree in the past), so many areas fall under the remit of more than one ITV company.

The current ITV contracts were awarded by the Independent Television Commission (ITC) in 1991 and give each franchise holder the right to broadcast for 20 years from 1 January 1993. There are 15 ITV companies serving 14 regions: Anglia Television (East Anglia), Border Television (The Borders and the Isle of Man), Carlton Television (London, weekdays), Central Television (East, West and South Midlands), Channel Television

(Channel Islands), Channel Three North East (North East England), Grampian Television (North of Scotland), Granada Television (North West England), HTV (Wales and the West of England), LWT – London Weekend Television – (London, weekends), Meridian Broadcasting (South and South East England), Scottish Television (Central Scotland), UTV – Ulster Television – (Northern Ireland), Westcountry Television (South West England) and Yorkshire Television (Yorkshire).

The ITV companies vary considerably in size and employment profiles. All provide programmes for the network and produce or commission regional programmes for their area audience. The larger companies, such as Carlton, Granada and LWT, broadcast to the most densely populated areas of the country and so have the highest income from advertising. They provide most of the network's drama and entertainment programming and therefore production jobs in any of these companies are likely to be highly specialised. The smallest companies tend to concentrate mainly on regional material and can offer excellent work experience because they are able to give their staff more involvement in a wider range of tasks.

Carlton, Meridian and Westcountry were originally established as 'publisher broadcasters', with the intention that they would commission the bulk of their programmes from independent producers. However, all have made inroads into original production, particularly Carlton, which now owns both Central and Westcountry (see page 62 for more details of the ownership of ITV companies).

Although training schemes within ITV companies are publicised, in general they are not widely advertised. Many contractors have established links with colleges in their franchise area, and may recruit graduate trainees from those institutions.

ITV Network Centre

The ITV Network Centre is a company wholly owned by all the ITV contractors. It is responsible for independently commissioning and scheduling all the programmes which are 'networked' on ITV (in other words, shown in every region at the same time). These programmes can be commissioned from independent production houses or from the ITV companies themselves. The Network Centre coordinates the coverage of major sporting events through its ITV Sport division, and it is also responsible for acquiring feature films and imported series for transmission on the network.

GMTV (Good Morning Television)

GMTV supplies a national breakfast programme to the entire ITV network, from 6am to 9.25am, 365 days a year. The station's news programming is provided by Reuters Television.

London News Network (LNN)

LNN provides local news programmes for the two London ITV stations, Carlton and LWT, which jointly own the company. LNN has a sports programme subsidiary, Independent Sports Network (ISN), which supplies sports services to GMTV, as well as to Carlton and LWT. Both LNN and ISN produce a small number of other documentaries and special programmes for wider audiences.

Channel 4 Television Corporation

Channel 4, which went on air in November 1982, provides a national service networked to the whole of the UK except Wales. The station offers a limited range of jobs because it only produces a minimal number of programmes itself; the vast majority are commissioned from independent production companies. Channel 4 employs 550 people; categories of staff include managers, engineers, commissioning, presentation, advertising sales, secretarial and computer staff. Numbers are expected to remain unchanged over the next five years. Most of those who work for Channel 4 are salaried employees. Freelance or fixed-term contract staff tend to be commissioning editors, continuity announcers or replacements for staff on holiday. Competition for posts is intense. For example, there are usually 100 applications for every specialised vacancy and 500 when a vacancy for an announcer's post is advertised. Channel 4 also receives around 350 unsolicited applications every month, a large proportion of which are for non-existent production jobs.

S4C (The Welsh Fourth Channel)

S4C schedules some 34 hours a week of Welsh language programmes supplied by outside bodies – BBC Wales, HTV and independent production companies. It also relays many of Channel 4's programmes, either simultaneously or on a rescheduled basis. S4C employs just over 100 people and recruits at all levels, but its intake of school-leavers is small. Staff turnover is also fairly low. Not all employees have to be Welsh-speaking, but knowledge of the language is essential in areas such as commissioning, presentation, and press and public relations. S4C is regulated by the S4C Authority, and not the Independent Television Commission (ITC).

Channel 5 Broadcasting Ltd

Channel 5, the UK's last national terrestrial network before the launch of digital television, started broadcasting on 30 March 1997. Eighty per cent of the UK's population can receive Channel 5 pictures. The company is a publisher broadcaster, so it commissions the bulk of its programmes from independent production companies or buys them from international markets. Channel 5 has 170 permanent staff, who are employed in areas such as programming, airtime sales, marketing, public relations, finance and business affairs. There are no plans to extend staff numbers beyond the present levels. However, the company does operate a graduate training scheme and also offers work experience placements to students of relevant subjects.

ITN (Independent Television News)

ITN is the company contracted, until the year 2003, to provide the ITV network with all its national and international news bulletins, such as News at Ten. The company also supplies bulletins to Channel 4 and Channel 5, and produces *ITN World News*, an international programme transmitted via satellite to overseas broadcasters and shown in-flight on several airlines. ITN also makes occasional one-off documentaries, and broadcasts live 'special event' programmes, such as budget and election coverage.

ITN Radio, a wholly-owned subsidiary company, is commissioned by Independent Radio News (IRN) to provide its news service, which is transmitted by over 180 regional and local radio stations. ITN is also part of a consortium which operates the London news radio stations, News Direct 97.3FM and LBC 1152AM.

ITN usually recruits experienced journalists only, although it will occasionally take on a small number of graduates with a demonstrable background in news for its Graduate Training Scheme.

Satellite television

The major player in the British satellite television industry is British Sky Broadcasting (BSkyB), whose themed channels are broadcast via the Astra satellite and are beamed directly to satellite dish-equipped homes across the UK and Ireland. Many cable operators also offer Sky's programming. Sky News, the 24-hour news and current affairs service, is the only Sky channel that remains unscrambled (that is, not part of a subscription package). BSkyB also has stakes in other satellite services, including the joint venture Granada Sky Broadcasting channels. Sky Television recruits staff at all levels, including junior secretarial and administrative posts suitable for school-leavers, but on the technical and production side usually takes on trained staff.

Satellite broadcasters tend to rely heavily on acquired material, although there are notable exceptions. Sky News is produced in-house, as is much of the coverage on Sky's sports channels. QVC, the home shopping service, broadcasts for 24 hours a day, of which 17 hours is live material. Most other channels either produce or commission at least a proportion of original programming.

BSkyB is also a partner in British Interactive Broadcasting, which is scheduled to launch a package of hundreds of channels and interactive services in 1998 using digital broadcasting technology.

A complete list of all licensed UK satellite broadcasters is available from the ITC.

Cable communications

Cable is one of the fastest growing sectors of the communications industry. Seventeen million homes across the UK now fall within areas franchised for cable, while most urban areas either have been, or are in the process of being 'cabled'. As a method of 'delivering' programmes, cable offers great potential to broadcasters; a single fibre-optic cable can simultaneously carry hundreds of television channels or 32,000 telephone calls.

British Sky Broadcasting is one of the main channel suppliers to UK cable operators. Some other channels are exclusive to cable viewers, although, as with satellite, original programming tends to be restricted by smaller budgets.

Cable companies employ 17,000 people in areas such as software engineering, technical design, sales, marketing and consumer services. Further information is available from the Cable Communications Association and the ITC.

Republic of Ireland

RTE – Radio Telefís Éireann – is the Republic of Ireland's national broadcasting organisation. Funded by both advertising revenue and a licence fee, RTE employs over 1900 people. It operates two national television channels: RTE1 and Network 2, the national radio networks RTE Radio 1, 2FM and FM3, plus an Irish language service, Raidió na Gaeltachta. In addition, RTE Radio Cork broadcasts to listeners in the province of Munster. TnaG (Teilífís na Gaeilge), a Gaelic television channel, was launched as a wholly-owned subsidiary in 1996, broadcasting programmes from RTE and the Irish independent sector. It will eventually become a stand-alone corporation, but at the moment, along with all other RTE services, it is regulated by the RTE Authority.

Other Irish broadcasters are regulated by Ireland's Independent Radio and Television Commission (IRTC). These non-RTE services, which are solely funded by advertising revenue, include 21 local radio stations, ten community radio stations, and the national broadcaster, Radio Ireland. TV3, a third English-language network, is planned.

Teletext

'Teletext' is a generic term which refers to the text information services broadcast alongside the regular television picture signals. Around 75 per cent of all television sets sold in Britain today are teletext-equipped.

The two major teletext operators in the UK are Teletext (with a capital 'T'), which is transmitted on ITV and Channel 4, and the BBC's Ceefax. Both operators provide a broad mix of information and features and update their services hundreds of times each day.

Most satellite and cable channels also provide a teletext service. Some are produced in-house, such as Skytext, which is broadcast on BSkyB's channels. Others are provided by specialist companies such as Intelfax, the UK's largest independent teletext operator.

Ceefax

Ceefax is produced by BBC News. It is available in the UK on both BBC channels whenever their transmitters are broadcasting, and in some parts of Europe. Around 2000 pages are on air at any one time, updated 365 days a year by a 24-hour newsroom at Television Centre.

Teletext Ltd

Teletext Ltd is based in London and employs roughly 120 full-time staff, with freelance writers hired as necessary. It operates 18 regional sub-services which provide local news, sport, weather, listings and arts coverage in addition to the national service. Like the ITV companies, Teletext is regulated by the ITC. Its current licence runs for ten years from 1 January 1993.

Independent television production companies

There are more than 1000 independent television production companies in the UK. The material they produce for broadcasting embraces everything from drama to current affairs and commercials. They also make music videos, corporate videos and feature films. Some companies employ their own full-time technical and production personnel, but the majority keep numbers to a minimum and hire freelance staff as and when required. Independent production companies supply the vast majority of British programmes on Channel 4 and Channel 5, while both ITV and BBC TV are legally obliged to commission at least 25 per cent of their output from independent producers. The Producers' Alliance for Cinema and Television (PACT) is the trade association for the independent sector.

Independent national radio

The 1990 Broadcasting Act made provision for three national commercial radio services, regulated by the Radio Authority. Each licence runs for eight years.

Classic FM

Classic FM, Britain's first independent national radio station, went on air in September 1992. It operates 24 hours a day and its output is based on classical music, with an average music-to-speech ratio of 75:25. Classic FM has a permanent staff of approximately 100 in areas such as programming, sales, finance and marketing. The station also has a reserve of freelances, whom it employs on a regular basis. Classic FM does not offer any sponsorships or work experience, nor does it run any training schemes. Its news service is provided by IRN.

Virgin Radio

Virgin Radio, launched in April 1993, plays a mix of new rock music and classic album tracks. It employs 60 people, including DJs, producers, accountants and sales and promotions staff. The station's news bulletins are supplied by Reuters. In April 1995, Virgin launched an FM 'simulcast' station, broadcasting to London and South East England.

Talk Radio

Talk Radio is Britain's first national commercial speech station. Phone-ins are a prominent feature of all programmes on the station, which launched in February 1995 and broadcasts on 1053 and 1089 AM. Talk Radio's news bulletins are supplied by IRN.

Atlantic 252

Although it broadcasts from County Meath in the Republic of Ireland, Atlantic 252 has targeted UK listeners since its launch in September 1989, and its long wave signal now covers more than 70 per cent of the country. Atlantic broadcasts 24 hours a day, playing current chart music plus selected 'mass appeal' tracks from the previous five years or so. The station is regulated by the RTE Authority, under an external Irish licence.

Independent Local Radio (IR)

There are currently some 160 stations in the IR network. Each supplies a commercially funded local radio service to a specific part of the country. Licences run for eight years and are awarded by the Radio Authority.

It is difficult to generalise about IR because, although several stations may be owned by the same group, most operate as independent companies and have a strong local character. Music forms the backbone of most IR schedules, while other sorts of programme are more specifically tailored to meet the needs and interests of the local population (for example, news, weather and travel information). Some stations also broadcast a small proportion of networked material.

Employment profiles vary from company to company, although every station is staffed by a mix of salaried employees and freelances. Categories of IR staff include programme controller, head of music, presenter, engineer, technician and sales, administrative and support personnel. Producers and production assistants in IR tend to be found only in the largest stations.

Vacancies within IR stations are not always advertised, as most will have a list of contacts on whom they can readily draw, with some people moving jobs between stations. IR stations also receive a huge number of unsolicited applications. Given the nature of the local radio network, IR companies show a strong preference for local applicants as they need staff who know the region's history, culture and social and economic situation – and who can pronounce local place names.

Training is an individual company responsibility, so again, policies vary from station to station. In general, though, there is an extremely limited intake of trainees, as stations prefer to recruit experienced personnel and give supplementary on-the-job training when it is needed. Several IR stations enjoy excellent relations with schools, colleges, Youth Training organisers and hospital radio, and provide work experience for young people.

A complete list of IR companies is available from the Radio Authority.

Regional IR services

Regional IR stations cover larger areas than traditional IR services and broadcast in addition to the existing companies in those regions. Regional stations are already broadcasting in Central Scotland, North East England, North West England, the Severn Estuary, East Midlands, West Midlands, Solent and the East of England, and further services are planned.

Commercial Radio Companies Association (CRCA)

The CRCA – formerly known as the Association of Independent Radio Companies – is the trade association of the independent radio industry. It represents the interests of member companies to the government, the Radio Authority and to other organisations.

Community Radio stations

These stations differ from those in the IR network in that each is locally owned and controlled and not run for commercial profit. The membership body of community radio is the Community Radio Association (CRA), which provides information, advice, training and consultancy services. The CRA also represents the interests of community radio to policy makers and other important bodies and holds regular conferences and events. There are currently 15 fully-licensed community-based radio stations in the UK.

Other employment areas

Facility houses

Facility houses hire out their technical services and equipment to other production companies. They are used mainly by the independent sector, although broadcasting organisations such as the BBC will also use them if suitable equipment is not available in-house, or a more competitive deal is being offered. Some facility houses will also provide crews to operate the equipment.

British Forces Broadcasting Service (BFBS)

BFBS is the broadcasting division of the Services Sound and Vision Corporation (SSVC), which is a non-profit making organisation funded by the Ministry of Defence. It provides the armed forces with a range of entertainment, engineering support and training services.

BFBS Television provides services in Germany, Gibraltar, Cyprus, the Falkland Islands and Bosnia. Its output consists mainly of programmes acquired from the BBC and the ITV companies.

BFBS Radio operates two radio stations: BFBS 1, a contemporary music channel which broadcasts 24 hours a day; and BFBS 2, a speech, information and light music channel, which carries programmes from BBC Radios 4 and 5 Live, in addition to its own original material. BFBS Radio broadcasts in Germany, Cyprus, Gibraltar, Brunei, Belize, Bosnia and the Falkland Islands. A taped service is supplied to isolated areas and Royal Navy ships at sea.

All BFBS staff are civilians. Approximately 200 people are employed worldwide by these services, including managers, administrators, producers, presenters, engineers, secretarial and support staff (such as drivers and messengers). The television and radio services mostly employ experienced, professional broadcasters. Both offer permanent, pensionable employment as well as short-term contracts.

Foreign broadcasting organisations

Many foreign broadcasters, such as the US networks, have offices in London. The key posts in these organisations will be held by foreign nationals, but it is possible that a number of support posts may be open to British subjects. An embassy should be able to supply the names and addresses of its country's broadcasting organisations that are running a London office.

Regulatory bodies

The Independent Television Commission (ITC)

The ITC licenses and regulates all commercially funded television in the UK (excluding S4C). Its remit covers the output of terrestrial broadcasters, such as ITV, Channel 4 and Channel 5, as well as that of satellite and cable companies. The Commission ensures that the overall mix of programme services caters for a wide range of tastes and interests.

The ITC has statutory powers to impose penalties on licensees if they do not comply with its various codes on programme content, advertising, sponsorship and technical standards. Penalties can also be imposed if licensees do not keep to their licence conditions. Compliance is monitored through the Commission's London headquarters, its regional offices and with the assistance of 11 regional Viewer Consultative Councils.

Radio Authority

The Radio Authority regulates and licenses the UK's independent radio industry. This includes all non-BBC local, community, cable, satellite and restricted services (such as 'special event' radio and hospital radio). It awards licences, plans frequencies and monitors programmes and advertisements. The Authority has the power to impose sanctions on any licensees that break its codes of practice.

Members of the Radio Authority are directly appointed by the Department of Culture, Media and Sports. The Authority's only source of income is from annual fees paid by the licensees.

Advertised vacancies

Vacancies in the organisations mentioned in this chapter are generally advertised in the trade press (for example, *Broadcast*), the media pages of the national broadsheets (for example, *The Guardian* and *The Independent* on Mondays) and, occasionally, local papers. Technical or specialist posts will often be advertised in the appropriate specialist publication. A summary of BBC vacancies is regularly updated on Ceefax, while Internet users can access information and advice about jobs, training and courses with the Corporation through their 'World of Opportunity' website. The address is http://www/bbc.co.uk.jobs

3 The jobs

A good rule of thumb is that the larger the company you work for, the more specialised your job is likely to be. In a smaller company, you will probably have a wider range of work. The broadcasting industry has witnessed a significant movement away from rigid job demarcation to multi-skilling and a shift from permanent contracts of employment towards freelance engagements. Most people who work in television and radio – both in front of and behind the camera or microphone – do not have permanent, pensionable jobs, but are contracted for a single programme or series.

Pay and conditions

It is difficult to give meaningful information on salaries because they vary from one company to another and change regularly. Many salaried staff receive considerably more than their basic pay each month because overtime, regional weightings, shift allowances and so forth are added on. Union members, of course, will have their rates of pay and conditions of employment negotiated by the unions and management. In general, all that can be said is that salaries in broadcasting compare favourably with those in other sectors.

The broadcasting trade unions

The role of the trade unions has changed significantly in recent years. Many collective bargaining arrangements have been weakened or terminated, and union membership is no longer essential for employment. However, the unions have found new ways of representing their members, for example in matters of contract law, insurance, taxation, debt recovery, health and safety, professional standards and vocational training. Many employers still regard union membership as a reliable indication of someone's skill and experience in the industry. The unions most closely associated with broadcasting are: The Broadcasting Entertainment Cinematograph & Theatre Union (BECTU); the Amalgamated Engineering and Electrical Union (AEEU); the British Actors Equity Association (Equity); the Musicians Union (MU); the National Union of Journalists (NUJ); and the Writers Guild of Great Britain. All these unions belong to the Federation of Entertainment Unions.

Job descriptions

The job descriptions that follow are presented in alphabetical order rather than divided into radio and television sections. This is because many posts are common to both media. The job titles given here will vary from one organisation to another, as will the precise areas of responsibility. It is also worth remembering that these descriptions are only indicative; none is an exact description of a particular post with a specific company. This is not an exhaustive list – the best advice is to use this chapter as a starting point for your own research.

Accountant

There are two kinds of work for accountants in broadcasting organisations. Financial accountants help to prepare and control long- and medium-term company and departmental budgets, provide financial reports and audits, supervise wage, salary and expense payments and manage pension funds. Programme accountants provide management with information on the state of programme budgets and what different departments are spending. Programme accountants need a thorough knowledge of how programmes are made in order to estimate and monitor production costs.

Actor (see Performer)

Agent

Many freelances working in broadcasting – including performers, writers and presenters – have an agent or manager to find them work, negotiate their contracts and fees, handle their publicity and help them to develop their careers. In return for these services, the agent takes a percentage (usually between 10 and 20 per cent) of the client's earnings. Agents need sound judgement, negotiating skills, legal expertise and business acumen. They are not employed directly by television and radio companies but work closely with them.

Airtime sales staff (commercial sector only)

Radio and television companies whose revenue comes from the sale of advertising time employ an in-house team to sell airtime or contract a specialist company to do the job. Sales coordinators negotiate the sale of airtime, take bookings and see that 'slots' are filled; sales (or marketing) executives are responsible for attracting new business; sales research staff carry out and interpret market research; traffic staff monitor the make-up of commercial breaks and arrange for the receipt and delivery of advertisements. Trainees for these posts are recruited externally, and advertising agency experience is beneficial. Television commercials are made by independent production companies in conjunction with advertising agencies, although many radio commercials are made in-house.

Announcer

Announcers, also known as continuity announcers, provide the links between and within programmes. They also project the image of the station, so different companies require very different people. Generally, though, announcers should have a well-modulated voice, a warm, friendly personality and, for television, an attractive appearance. Their work usually involves writing all their own continuity scripts, but it can also involve interviewing, reading scripted commentaries, news bulletins and even short advertisements. Most announcers have had an A-level or higher education and, increasingly, previous media experience. They frequently work from self-operated studios.

Case Study

Gillian *is an announcer with an ITV company. Her duties include linking the programmes on the channel (at 'programme junctions') and reading some news and weather bulletins.*

'Typically, I would start at 10.30 in the morning, and go through until 7.00 in the evening. The only drawback is that you don't get any lunch or tea breaks – you have to be there all the time in case there is a problem with any of the programmes.

I read the news at 10.50, so I have roughly 20 minutes to gear myself up for that. It doesn't matter how long the bulletin is, the pressure is the same. I also have to prepare the weather graphics which are sent to us via satellite from Birmingham and make sure that the script isn't full of meteorological terms; it's up to me to make it as understandable as possible.

This job can be very stressful. Sometimes you can be put on the spot – for example, if there's a switching problem with the lines for an incoming programme, like *News at Ten*, and it 'goes down', then it's up to the announcer to ad lib and fill that gap.

So, I would say that to do this job, you need to be fairly sharp, you need to be on the ball. You also need to be very disciplined. You must be here to fill the slots that are assigned to you, because nobody else is going to do it. Also, you can't just lie down when something goes wrong or you make a mistake. You've just got to hope that it's going to be better next time. I always remember being told that you're only as good as your last junction, and it's absolutely true!

A lot of people perceive this to be a glamorous job and it's not really. You have to be realistic about that. One of the worst things, I suppose, is that we suffer because of other people's mistakes. Although I read some news bulletins, I am not a journalist; I have no input into the news whatsoever, I am just the messenger. If, for example, I haven't had time to read through the autocue before a news bulletin and a journalist has made a mistake in a news script, then it would be up to me to try and cover for them. If that doesn't come off, I am the one left looking and feeling rather stupid. Having said that, when you're having a good day, it's the best job in the world.'

Audio work (see *Sound operator*)

Camera operator

Two kinds of camera are used in television: electronic (or video) cameras, which record on videotape; and film cameras, which are mainly used for commercials and expensive dramas. The work of a camera operator is tiring. It involves a lot of standing under hot studio lights, the hours can be unsocial, and the equipment can be very heavy. Camera crews are often required to work away from base for long periods, sometimes in uncomfortable conditions. Recent developments in technology and in the workplace, such as multi-skilling, have particularly affected the work of TV news camera operators, many of whom have required basic retraining.

Camera operators are usually recruited at the age of 18-plus and trained on the job. Applicants should have a good general education (including GCSEs in maths, English and a science) and possess some knowledge of optics, film and television photography. They must also have normal colour vision and a good eye for picture composition. Some companies start their trainees operating cameras early on, but others make them spend several months operating cranes and moving cables. It usually takes six years to become fully trained, and promotion to the top grades of the profession can take between ten and 20 years.

Video camera operator
Experienced video camera operators may decide the framing and the composition of the picture, but will be guided by the

programme director on how the action is to be shot. Those working with few other people or alone have to decide for themselves how to achieve the most telling shots. They must also be able to service their equipment, edit tapes and use electronic communications links equipment.

Film camera operator/lighting camera operator

Senior film camera operators, also known as lighting camera operators, are responsible for the artistic and technical quality of the pictures and, with the director, share major decisions about camera positioning, lighting and how the action will be shot. Trainees begin by performing tasks such as pulling focus, loading and charging magazines, checking and cleaning equipment and recording takes.

Casting director

Casting directors, many of whom are freelance, work with the producer and director on the casting of a production. They liaise with agents, visit drama schools and spend a lot of time interviewing and auditioning hopefuls. Casting directors need to have an excellent memory for faces and an instinctive knowledge of who will be right for a certain part. In many radio and television productions, however, the casting will be done by the director and/or producer.

Continuity announcer (see Announcer)

Costume designer

Costume designers work in the areas of television light entertainment and drama. They begin by reading the programme script then, in liaison with the producer, director, choreographer and set designer, plan the costumes and work out the costume budget. Costume designers should have a good grounding in the history of costume and etiquette and be creative and innovative, while possessing administrative and supervisory skills.

Costume design assistant

This is one of the entry points into the profession. The assistant's duties include researching, arranging fittings and shopping for fabrics. Applicants normally hold a degree or equivalent qualification in theatrical or fashion design.

Designer (see *Set designer*)

Director (see *Programme director; Casting director*)

Disc jockey/DJ (see *Music presenter*)

Dresser

Dressers are responsible for the maintenance of costumes and for helping artists on and off with their costumes. They carry out minor alterations and must therefore be able to sew quickly and neatly. Essential personal qualities are tact, maturity and sensitivity. Both men and women can work as dressers, but people under the age of 20 are rarely recruited for this work.

Editor – videotape, film

The editor, working closely with the producer and director, prepares the final version of a programme. The work demands great attention to detail, precision and creativity, and the skills take a long time to accumulate. A relaxed and modest personality is also an asset because, ironically, the better the editor, the less conscious the viewer will be of his or her work. Most television programmes are recorded on videotape cassettes. When edit points have been decided, the sections of the tape to be used in the production are recorded on to another tape. Film, however, is physically cut and spliced. Increasingly, editors are becoming multi-skilled and able to work on all formats. Technological developments, such as non-linear editing, have effectively eliminated the traditional entry-level grade of assistant editor.

Engineer

Engineers have a vital role to play in broadcasting, but it is beyond the scope of this book to give details of all the jobs available in this vast and highly technical field. Further information can be obtained from BBC Recruitment Services.

Film camera operator (see Camera operator)

Film editor (see Editor)

Floor manager/stage manager

Floor managers coordinate and manage everything that happens on the studio, location or rehearsal room 'floor', including checking that props are in place and making sure that performers know where to stand and what to do. They also give cues and prompts and, if there is one, take charge of the studio audience. Floor managers usually start their careers as assistants and need to be thoroughly familiar with every aspect of television production. Theatre experience is also helpful.

Graphic designer

The graphic designer is responsible for designing and supervising all graphic programme material, including credits, charts, graphs, logos and even some props, such as documents used in a drama. The graphic designer must be highly creative, but also capable of adapting concepts to suit the requirements of production staff. They work in close consultation with the programme's producer, director, set designer and sometimes, when developing a title sequence, the composer of the theme music.

Case Study

Liam is a graphic designer with a regional television company.

'I've been here 18 years now and have seen a lot of changes in that time. Advances in technology have made it much easier to achieve "glossier" looking results, but that just means that overall standards are now higher, so that adds to the pressure.

In fact, there are a lot of pressures in this job, such as trying to get a graphic finished for the nightly news programme. Deadlines have to be met. Also, producers will often come to us with preconceived ideas of the effects they want us to achieve, but they may not be possible on our equipment. Sometimes they just want you to duplicate what they've seen elsewhere. So, in effect, I'm trying to compete with someone else's ideas.

On the plus side, there is tremendous satisfaction when you've completed a piece of work that you are pleased with – although more often than not the producer will then come in and pick holes in it! But that's just part of the job and you have to get used to it.

I think the best graphic designers are the people who can come up with an unusual slant on things, or who can instantly see a clever or novel approach and make it work. Unfortunately, nowadays the effects you can achieve tend to be down to how much someone is prepared to pay for a particular job. That's probably what annoys me more than anything else – ten years ago money wasn't the dominant factor, whereas now budget pressures are really stifling creativity.'

Journalism and news work

News is a very important part of any broadcasting organisation's output. Many people are involved in news gathering, writing and presentation, but everyone from the news typist to the foreign correspondent works under the same intense pressure. There are strict deadlines, bulletins are always broadcast live and stories must be constantly updated. News work never stops and, with early morning and late night news bulletins, it involves unsocial hours, weekend duties and night shifts.

News reader

Also called news presenters, newscasters and anchormen/women, news readers are frequently experienced journalists. They present the news from the studio, linking and introducing stories from other journalists. They may read from a prepared script or one that they have written themselves. They must be thoroughly acquainted with the background to the news and must check things such as the pronunciation of foreign names.

A news reader must be able to read at sight without fluffing (in case a story breaks in the middle of a bulletin) and to vary the pace of their delivery to fit with time signals and call signs.

Journalist

There is fierce competition for journalists' posts in broadcasting. Applicants for traineeships generally hold a good degree or an equivalent qualification, and all can show examples of their best work (in newspaper articles or on audio or videotape). All journalists or reporters must be able to type and take shorthand.

Case Study

Mark *is a Senior Broadcast Journalist with a BBC national region.*

'I started here doing freelance radio reporter shifts. I later moved on to produce some of the radio current affairs programmes before I was offered a staff contract. I don't enjoy producing as much as reporting, but with multi-skilling nowadays, you don't really have a choice!

After I'd been doing the job for a few years, I started to get a bit bored; I just felt I needed a change. I also fancied a stab at working in national radio, and luckily, I was able to arrange a swap with one of the reporters from *The Breakfast Programme* on Radio 5 Live.

I found things very different working on a national station, compared to a regional one. For a start, I had to have a much wider field of reference to do the job, as I was now reporting on national issues, and that was a bit daunting at first. Radio 5 was also a lot tighter on the production side. Every aspect of production was completely mapped out to the last detail; some of it seemed a bit obsessive to me at first, but the upshot of it all was that nothing ever went wrong!

I'm back here now, but on attachment in the television newsroom. Writing television news reports is definitely a different skill to writing for radio. Say you're covering a story about the devastation caused by a bomb; with television, the image itself is powerful enough to tell the story whereas, with radio, your words have to paint the picture for the listener. You also don't get as much time in television, but it's arguable that the pictures allow you to pack in the same amount of information. That said, I don't think the journalism is as good in TV as it is in radio. There are honourable exceptions, of course, but the television medium is so picture-led that it's much easier to be sensationalist.

To do this job you need a curious nature, a lot of front, and, at first, you have to be happy working for long hours for not very much money. But possibly the most important skill you need is judgement – knowing what to leave in and what to leave out. A good voice is also important and, if you're in television, so is a reasonable appearance. That may seem shallow, but it's the nature of the business.'

Case Study

Adrian *is a freelance journalist, with experience in reporting and presenting on both TV and radio.*

'You have to get used to the idea that your diary can be full one week, but empty the next. That's not a huge problem, because a lot of my work comes at very short notice, sometimes on the day itself – maybe I'll be called in because someone is sick, or whatever. But then sometimes you find that the diary is empty and it stays empty. I've had bad periods where I've gone for a month or more without working at all.

You must always be on the lookout for story ideas, and always think about who you can sell them to. You may even be able to sell the same basic story to different programmes – providing, of course that you take a different angle and don't use the same interviews.

When you're freelance, you have to get to know all the producers in each department of the various companies. There is a constant and developing relationship, and there has to be a lot of trust on both sides. You will hear of cases where producers steal a freelance's idea and give it to a staff journalist, but personally I have never been burnt in that way.

It does help enormously if you have your own equipment – tape recorder, mobile phone, or whatever. A big advantage about being freelance is that, for tax purposes, you are self-employed, so these expenses can be claimed back off tax.

It's important that you structure your day so that it doesn't get out of control. If I'm reading the regional news on GMTV, for instance, I'll work from 4am until 9am. If I've been offered a day shift with the BBC, I'll start work there at, say, 9.30am. If somebody then offers me evening work – and it's not always easy to say no, because you're always afraid they won't ask again – you end up burning the candle at both ends, and believe me, it gets to you very quickly!

The great thing about being freelance is that you are your own boss – although only to the extent that other people ask you to work for them. You also have total freedom to decide when you take your own holidays. Unfortunately, people invariably want you to work for them during the best holiday times, such as Christmas and New Year!'

Librarian

Librarians in a broadcasting organisation are more likely to handle
'non-book' material, such as film, videotape, scripts, music
scores and newspaper cuttings, and will be expected to have had
appropriate training. There are a number of posts for graduate
chartered librarians, assistant librarians and clerical library assistants. Librarians perform the normal range of library tasks and
will, in some cases, also carry out research. Film and videotape
librarians need a knowledge of production, storage and handling
techniques and can be asked to do simple editing. Music librarians in a small organisation may need a wide knowledge of music,
but in a large organisation their work is likely to be specialised.

Lighting camera operator (see Camera operator)

Lighting director

Lighting directors decide how to position lights in order to
produce the best effect on a given set. They prepare plans for,
and supervise the work of, the lighting electricians and lighting
console operators. Lighting directors must liaise with the programme director, the set designer and other members of the
production team, such as the make-up artists. They require both
flair and technical ability and are usually recruited internally.

Lighting electrician
Lighting electricians follow the plans of the lighting director,
arranging studio lamps on the lighting grid in the studio roof so
that the desired effect is achieved. They also repair and maintain
the apparatus they use, so a good head for heights is essential.

Location manager

The services of a location manager are needed most frequently
in connection with drama productions. If, for example, the script
calls for a Jacobean house, it is the job of the location manager

to find a suitable one, obtain permission to use it, arrange for such things as parking space, and see that the place is left in good condition when the shooting is over.

Make-up artist

Make-up artists spend much of their time doing corrective work, such as combing hair and powdering noses, or carrying out routine tasks such as setting up and stocking equipment. The creative aspect of their job comes to the fore during drama and light entertainment productions, for which they may be required to do elaborate face and body make-up, style hair and wigs or produce effects such as scars and bruises. Make-up artists need a calm, tactful personality as they work with many different sorts of people – actors, politicians, ordinary members of the public – all of whom are likely to be nervous before going on camera. Make-up artists usually start as trainees. General requirements include normal colour vision, a good general education (A-levels in English, art and history are particularly desirable) and recognised qualifications in make-up and/or hairdressing, and/or beauty culture. Art school qualifications are sometimes acceptable.

Meteorologist

Weather forecasts are generally prepared and presented by qualified meteorologists, who see radio and television work as only a small part of their job. Some companies specialise in the production of television weather bulletins, such as International Weather Productions (which prepares the forecasts for GMTV and those which follow ITN's news programmes on ITV) and The Weather Department (responsible for the bulletins on many of the regional ITV stations). Both companies employ very few staff, all of whom are highly trained and experienced, and are therefore not in a position to offer work experience.

Music presenter/DJ

This is one of the most sought-after jobs in radio, and stations receive hundreds of unsolicited applications and demo tapes. There is considerable responsibility attached to the position, as the DJ's personality represents the image of the station. A music presenter has to be a very good all-rounder; in addition to technical experience (music presenters in local radio operate their own studio equipment), the job calls for an easy microphone manner, a grasp of current affairs, a pleasant voice, and the ability to think quickly and to ad lib when necessary.

Case Study

Joe is a music presenter with an independent local radio station.

'My first experience of presenting on radio was in a clothes shop of all places! I was working there on a training scheme about eight years ago, when the guy who was the DJ for their in-store radio service took sick. So, having a good head for music, I volunteered to give it a go – and ended up doing the job for 18 months. I just loved it, I took to the whole thing like a duck to water.

It was a great place to learn the basics of communicating with an unseen audience, and I also found it a good way to learn how to convey a "radio" personality. When I first started, I tried to be too jokey, copying the likes of Steve Wright, but that just sounded very phoney. Everyone has their own niche, their own style, and it's pointless trying to imitate somebody else.

After that, I spent a couple of years with a community radio station before I was offered a job here, at one of the major local independent radio stations.

I have virtually no say in the music played on my programme, and that's probably the case with all music-based local radio stations. Everything, with the exception of maybe one or two tracks, is ultimately determined by our head of music who decides what goes on the playlist. The number of discs we play in an hour depends on how many commercials have to be broadcast.

In music radio, you really don't have an awful lot of time to stamp your personality onto a programme. You'll maybe have a 15-minute block of music, with 30 or 40 seconds of yourself speaking before the next block starts, so you need to have a fair idea of what you're going to talk about. A lot of people try to get away with just winging it, and most times it fails miserably. So I constantly scan the papers and watch a lot of TV for ideas for the show. Even when I'm at home, I never stop looking for material.'

News work – (see *Journalism*)

On-screen promotions staff

The promotions department is responsible for the on-screen presentation and promotion of a television station and its programmes. BBC Television promotions are coordinated in London, whereas each ITV company has its own presentation and promotion department, although these vary greatly in both size and technical capability.

Case Study

Alison *is a promotions assistant with a small ITV company. After completing a design degree, she worked as an advertising copywriter and art director before landing her present job.*

'I work alongside the Head of Promotions, who has overall responsibility, and the Promotions Scheduler, who decides which promos will be shown in which slots. We are a small team – most ITV companies do have larger promotions departments – but then we make fewer programmes than most other companies.

We are responsible for any on-screen promotions that are broadcast. For example, you might have three minutes of airtime between programmes, two of which could be taken up by commercials. That then leaves a minute which we would fill up with our own promotional material.

All the ITV companies link up by phone once a week, when we tell each other which programmes we are putting out and what promotional material is available for those programmes. We will then make a list of all the material we can access, and from that we'll decide which programmes need most support. Then we'll get a copy of the relevant material and edit it in such a way that we have our own 30-second commercial for each programme.

I would say that to work in promotions, you have to be creative, but you also have to be adaptable with your ideas; and you need to be able to take the most convoluted idea and turn it round very simply. You need to know what looks good on screen, and it also helps to have a good ear for music.

Really, what we are doing is advertising programmes instead of products, with the ultimate aim of boosting viewing figures because more viewers ultimately make a television company more money. All we can ever hope to do in promotions is to generate interest and to get people to tune into a programme. They may, of course, turn over after five minutes, but if we have got them to the point of tuning in for it, then we have done our job.'

Performer

Performers (actors, dancers, musicians, etc) work on a contractual basis. Most of them hire an agent or manager to find them work and negotiate their contracts.

Case Study

Tom O'Connor, comedian and quiz-show host, offers this advice to anyone contemplating a career in the entertainment industry.

'Show business is not all glitter and fairy lights. In fact, the opposite is true. Hard work, determination and, most of all, the ability to listen to criticism, are the only ways success is finally achieved.

I suppose the years of grafting, experimenting, dying the death and finally succeeding leave the entertainer with a feeling of permanent invincibility. A case of "It was a hard ride, but now I'm in the Promised Land – it's time to lean back and reap the rewards." This is the most dangerous time in anyone's career. The top is not the end, it is only the beginning and at this point an act needs lots of help, lots of encouragement and lots of honest people around.

Family and their support are vital. Managers need to be reliable, reputable and totally in your corner. Unfortunately, I've been the victim of one or two unscrupulous members of the profession and suffered loss of work and money because of their scheming and cut-throat dealings. A good manager, as I have now, can not only lengthen a career, but broaden it. A bad manager, by the very nature of the beast, can not only hamper improvement, but can seriously damage the good work already done.

It's easy to speak of caution from what may appear to be a lofty position, but, believe me, I only wish to help, and prevent the problems which have beset many like me.

Listen to family, friends and all who wish to criticise constructively. Watch all dealings done in your name, read all contracts before they're signed and keep a copy. Never assume that all is well even when it appears so – be suspicious. "Talk softly and carry a big stick", as they say.

And always remember the first rule of show business: when in doubt – don't!'

Actor
Acting is a notoriously overcrowded profession. In a recent survey, Equity, the actors' union, found that only 20 per cent of

its 36,000 members had worked for 30 weeks or more in the previous year, and that 25 per cent had not worked at all in that time.

Presenter

Presenters 'front' or host non-fiction programmes, from current affairs to quiz games and chat shows. Presenting jobs are rarely advertised; normally, a presenter will be approached (through his or her agent) by a producer and offered a contract for a specific programme or series. Many people have broken into presenting by becoming well-known in another field, such as sport or politics for example. It is very difficult to generalise about the specific qualities a presenter needs because these will depend totally on the nature of each programme.

Case Study

Mike Edgar has presented a range of programmes on both television and radio for BBC Northern Ireland, including the award-winning youth programme, Across the Line.

'Television presenting and radio presenting are two very different disciplines. Radio makes you think a lot more about what you are doing and what your guest is saying to you; *you* must make it sound interesting, *you* must paint the picture and draw the very best out of your guest, because you can't fall back on graphics or fancy camera angles to make it look interesting.

I much prefer radio as a medium, although I have had loads of good experiences with TV. The real beauty of television is the exposure and that is a marvellous thing, because it ups your value as a commodity, which is really what presenters are in this industry.

I find that the trouble with television is that too many people have a hand in the final product. Researchers, assistant producers, producers, the director, the executive producer, all of them feel, quite rightly, that they should have an input into what you look like, what you should say and how you should say it. With radio, you can be a lot more creative, a lot more hands-on than you can ever be with television.

When you are doing interviews, you need to be genuinely interested in the subject matter, even if you have been given a subject that is not within your normal working arena. You must do your homework. There is nothing worse than watching or listening to an interview and it becoming painfully apparent that the presenter hasn't read the interviewee's new book, or whatever. It is also very important to put your guests at ease straight away, and you must treat everybody, young or old, with respect.

Over the years, I have had the opportunity to present a lot of television and radio programmes. I have had a very interesting and varied time, I've met a lot of wonderful people, I've travelled a lot and I've been paid to do it. I can't believe my luck!'

Producer

Producers turn ideas into programmes. They head the production team and are responsible for managing the programme budget and the scheduling of rehearsal and recording time. They also have a say in the selection of programme participants. In radio, the producer often originates and researches the initial idea and records the material, although these duties can be shared with a researcher and production assistant. In television, the producer will often have conceived or contributed to the idea on which the programme is based. In some cases, the producer will also direct.

Producers must be receptive to the ideas of others and be able to originate suitable programme concepts. They also need to understand the requirements of the network or region in which they are working. Every programme has a producer and most producers specialise; for example, a current affairs programme needs a producer with a good grasp of the subject and sound political judgement.

Case Study

Owen is a producer with a regional BBC radio station.

'I started off in the BBC doing reviews of a local arts festival for some radio programmes when I was at university. They wanted the opinions of a student layabout to counterbalance the professional critics! Once I had my foot in the door I just carried on really, doing one review a week of some of the things going on about town. At about the same time, I started making packages and reviewing gigs for a nightly youth radio programme.

Out of the blue, the producer of that programme decided to take an attachment to London and I was invited to apply for his position on a short-term contract. I honestly didn't think I was eligible, but I found I had learnt a lot just by watching people, and I had no problem setting up interviews or choosing music. So, to my surprise, I was contracted for a month, then that was extended and extended until I eventually ended up producing that programme for about a year.

Then I moved on to help launch a new morning radio discussion programme. In a sense, this wasn't as easy to put together because it was a speech-based programme so we couldn't hide behind records. The show was built around a very strong presenter, so all the material had to be geared around his own personal likes and dislikes; it was a question of getting to know him and winning his confidence.

I then moved into television and spent the next year working as an assistant producer on a weekly chat show. I have to say that I personally didn't find it as fulfilling as my radio work. Radio is much more hands-on and I missed that very much. With television, you are working with a much larger group of people and there isn't the same scope to be creative.

So I made a conscious decision to get back into radio and I'm very happy back here now, producing documentaries. I really enjoy working as part of a small team. Ideas are shared, and watching an idea grow into a series is a great process to be involved in. But I've no doubt that I'll be back in live radio one day. I think it is essential to keep your hand in in all areas because these days a producer is increasingly expected to be a jack-of-all-trades.'

Case Study

David, who runs his own independent television production company, explains some of the issues involved in working in the independent sector.

'Being an independent producer means you have to pitch your ideas to the broadcasters, so one of the things you have got to do is watch television and read the media pages in the national papers to find out what the agendas are at the different TV stations.

You also have to get to know the people who commission in the different TV stations, find out what they like and what they don't like and try to make your ideas fit with their needs. A lot of people don't do that and are not successful. You also have to be original and come up with good ideas that work, and then when you have a good idea make sure you have the people who can turn that good idea into good television.

You also need an awful lot of good luck. There are people who have been independent producers for 20 years and are still struggling to get commissions. Yes, it is hard to get into television, but the fact that you are in does not necessarily mean that it gets any easier, even though it may seem that way looking in from the outside. For an awful lot of people in television, it is month-by-month, year-by-year survival.'

Production assistant (PA)

Production assistants work on a particular programme from start to finish, providing support services for the producer and direc-tor. In television, the PA sits with the director in the control room and passes instructions on to studio staff over the talkback system. In both television and radio productions, the production assistant times the recordings or takes with a stopwatch. They also provide any information needed by those engaged in post-production work, such as editing. Radio PAs tend to be found only in the BBC and larger IR stations. The job requires excel-lent organisational skills, attention to detail, calm, initiative and the ability to manage and get on with all types of people. Most PAs are recruited as trainees, although many vacancies for traineeships are filled by people already in broadcasting. Many applicants are graduates with secretarial experience.

Case Study

Joanna is a production assistant with a regional BBC radio station.

'The exact nature of what I do depends totally on the type of programme I'm working on; a comedy or documentary will have different requirements to a daily music programme, for example. In general, though, my job involves liaising with guests, typing up scripts or running orders before we head off to the studio and, once a programme has been transmitted, completing a Production as Broadcast form. This is a summary of everything that went into the programme – for example, how long each interview lasted, whether or not guests have to be paid and how much. I'll then send the relevant details down to the contracts department so they, in turn, can pay the guest. It's my job to keep track of all programme expenditure – artist fees, taxi bills, competition prizes – and deal with any other invoices.

We have to time every piece of music we play to the exact second, and also record details of the composer, the artist, the publisher, and what album it was taken from, so that everyone can receive their royalty payments.

A radio PA often has to work on more than one programme at once, so it's vital that you learn how to prioritise and manage your workload. You need excellent typing skills and, because you're often dealing with the public – whether it's on the phone, by letter or meeting someone in reception – you've also got to be extremely patient and polite! You need to be able to stay calm under pressure, because all sorts of deadlines have to be met. You have to be able to work on your own initiative, because the PA will often be left in charge of the production office.

A couple of years ago, I took a training attachment as a TV production assistant. I must say, I found television a lot more hair-raising than radio – suddenly I wasn't working with just one or two other people, but about 25! It was so much more involved and a lot more stressful. In fact, that's probably the main reason I decided to return to radio – I felt my personality just wasn't suited to such a nerve-racking environment!'

Production manager

The production manager organises all essential support facilities for a programme's production team. Working closely with the producer, their work involves negotiating contracts with broadcasters or clients, ensuring that legal contracts for staff and locations are drawn up, and that insurance is in place. The production manager arranges all the logistics of a shoot, and is responsible for the health, safety and general welfare of everyone working on the programme. Throughout, they must also ensure that all departments are working on schedule and within the allocated budget.

Case Study

Helen is a freelance production manager.

'The role of the production manager varies considerably depending on the nature of the production and the size of the budget. But on any scale, you have to be a good team motivator and have fantastic organisational skills because you are the key communicator. You also need to be able to grasp immediately the overall requirements of the production, as well as having a sense of humour and buckets of common sense to help solve all the problems that will arise.

The ultimate goal of a production manager's job is to ensure the smooth production of a programme – but the programme is, obviously, for a company which has to make its forecast profit. So it's essential that you juggle the budget within the perimeters of the production's demands.'

Programme director

In television, the programme director is in charge of the shooting of a programme and the direction of performers and technical crew. TV drama productions involve detailed shot-by-shot work. The director places the actors and discusses their interpretation of the roles and, after consultation with the technical experts, decides on lighting effects, camera angles and so on. When shooting has been completed, the director supervises post-production work, such as editing and sound dubbing. In radio drama, the director rehearses the actors, selects sound effects and, again, supervises post-production. In a live broadcast such as a news bulletin, the programme director follows a running order, selects pictures from those offered by the camera operators or from videotape, and relays instructions to the presenters. Most vacancies are filled by internal applicants with substantial production experience.

Promotions (see On-screen promotions staff)

Property buyer

Property/production buyers buy or hire movable items or 'props' for a studio set. They may have to supply potted plants for a chat show studio, food for a sit-com or furniture for a period drama, for example. Nearly all property buyers have had theatrical experience or have worked their way up the props department.

Reporter (see Journalism)

Researcher

Behind every successful programme there will have been a hard-working, competent researcher. The exact job description will vary from programme to programme, but their duties can include lining up and interviewing contacts, conducting vox pops, finding locations, digging through statistical information, checking through archives, and scriptwriting. They are also expected to contribute ideas to the programme. Researchers can specialise; for example, some companies employ contestant researchers specifically to find suitable contestants for their quiz and game shows.

Case Study

Elaine *is a freelance researcher. This is an extract from her work diary to illustrate a typical day's research for a television documentary on the subject of teenage pregnancies.*

'*In by 9am:* Spend an hour phoning people, either introducing yourself, persuading them to help you, or getting permission to do something, for example, film an ante-natal class. Write letters and leave them to be typed up.

10am: Travel to first meeting: a pregnant 17-year-old living in a hostel. Aim: to get more information on prepared subjects and ask her if she is willing to take a feature role in the programme, including the filming of her scan. She has previously requested anonymity. Try to talk to another mother who is due to have her 6-week-old child taken away from her for adoption. Request permission from the hostel manager to film on the premises.

12 noon: Back in the office. Check the post and scan the papers for any relevant info. Respond to any messages left. Write up notes from last interview while having lunch.

1pm: Travel to next meeting. Second meeting with a specialist and his researcher who may be persuaded to give you access to their questionnaire and database. Enquire about relevant reports and articles.

2.30pm: On to next interview with a couple of young mothers in a rural town and their volunteer worker. Persuade them to let you record the interview – that makes it a long, detailed meeting.

5pm: Back in the office, check out references with the library and go and get them if possible. Call into the government bookshop on the way and make a list of the publications you want permission to buy. (I had to return and spend over an hour copying the stats and references down, because we couldn't afford to buy them all.) Otherwise, get back on the phone, often sifting through specialised directories or ringing press officers. Check letters and send them.

Things to do daily: Add to your biography lists and update the producer.

Make lists of things to do tomorrow and write interview questions for next meeting.

Homework: Go out to evening meetings two or three times a week. Read and summarise any reports you have been lucky enough to get hold of. Prepare summary reports for weekly production meetings.'

Elaine adds: 'In a sense, this was only half a day's work, because I was actually working on two programmes at once. Finding the right ideas took about four weeks of a 12-week contract and finding the right people took about another six.

The most important factor is timescale. With recording dates set well in advance, you must meet deadlines. This means organising your time, keeping a production book with every scrap of information and, most importantly, being motivated to work on your own initiative. After all, it could be your responsibility to fill every second of airtime.

Gathering information is only part of it. Clear and concise presentation skills are needed to sell your ideas to the production team. There's no point in having discovered a critical argument unless you can make it come alive. You must have ideas and information to contribute. This resourceful attitude is the key to the job.'

Runner

Runners act as general assistants on a production. Their duties can include delivering packages, photocopying scripts, taking messages – and making lots of coffee! Runners are often highly qualified; many people are so eager to break into broadcasting that they are prepared to apply for posts that they would not consider accepting in any other sector, in the hope that once inside they will be able to work their way up the career ladder. This is the traditional entry-level job for the industry, and although the financial rewards are small, it is recognised as an excellent way to learn the business from the inside.

Sales staff (see *Airtime sales staff*)

Script editor

Script editors work in drama departments in close consultation with programme producers. Their duties can include commissioning writers, finding new writing talent, conducting research, and rewriting material. In the case of a long-running serial, where a number of writers are needed to contribute different episodes, sometimes over a number of years, the script editor ensures that each is written in a uniform style and that plot and character details remain consistent. The script editor may also be responsible for developing storylines – coming up with plot ideas in consultation with the producer and the other writers, and deciding how much of the plot should be revealed in each episode.

Secretarial work

Although secretarial work in a broadcasting organisation can provide a rewarding career, many people see these posts as a springboard to more demanding senior jobs, particularly in production, and for this reason they attract a high proportion of graduate applicants. However, competition for production posts is extremely tough and many require specialised knowledge and experience – and secretarial work is not necessarily the best way to gain that experience. In addition, all staff are expected to remain in their first post for a reasonable length of time before applying for a transfer.

Set designer

Every television programme, from a game show to a drama, has a set which has been designed to create a certain ambience. Designers work with producers and directors and must have a feel for how the content of the programme should be visually interpreted. The work requires knowledge of the history of art and architecture. Studio sets are viewed in close-up and from different angles, so a great deal of planning goes into them. Designers have to take into account the positioning and move-ment of performers, props, cameras, camera cables, microphone booms and lighting. Much design work is now done on com-puter, while recent years have also seen the development of 'virtual sets' – studio backdrops which don't physically exist, but are designed, generated and stored on computer. Set designers almost invariably begin as assistant designers, having obtained an art college degree (or its recognised equivalent) in interior design, art and design, stage design or architecture.

Setting assistant (see Stagehand)

Sound operator

Sound operators require both creativity and technical skill. In a radio studio, their duties can include sound balancing, mixing and recording. However, the introduction of digital editing and transmission systems, which do not require the specialist skills of sound operators, has led to job cuts in this field; for example, computers now control the output of many independent radio stations, particularly during the night-time hours.

In television, sound operators must also ensure that the studio equipment is functioning properly and, working with the camera operators, they see that microphone boom arms are effectively positioned and yet remain out of shot. During post-production, they may be involved in recording, editing and dubbing taped speech and music, as well as selecting (and, in some cases, devising) sound effects. On outside broadcasts, such as coverage of a live football match, sound operators are responsible for rigging up and dismantling equipment and for testing the quality of the lines to studio.

Sound mixer
Sound mixers monitor and balance the sound signals from a control room. They also feed in music or other sound effects where appropriate.

Special effects designer (see Visual effects designer)

Sports reporter/commentator

Sports reporters are normally journalists who specialise in sports coverage, and many commentators are retired sportsmen/women. A good microphone manner is essential, as is a thorough knowledge of the sports they cover. Commentators need to be able to speak fluently and coherently without a script.

Stagehand

Stagehands, also known as 'setting assistants', usually work in small teams. They erect the scenery in a studio or on location and dismantle it when shooting or transmission is finished. Safety is crucial. Generally, stagehands have had craft training through the theatre or construction industry.

Stage manager (see Floor manager)

Transmission controller

Transmission controllers are responsible for sending a company's programmes to the transmitters so that they are broadcast at the advertised times. They monitor the pictures currently being broadcast and those which are about to be sent to the transmitter. The transmission controller must ensure that each programme is brought in on cue. The work involves a great deal of planning and combines periods of intense activity with periods during which very little happens. Transmission controllers need to be able to think and act quickly in the event of a technical failure as viewers must not be left with a blank screen. Most vacancies are filled by internal applicants.

Vision mixer

Television programmes are made up of pictures that come from a number of different sources, such as a camera in the studio, or pre-recorded videotape. Vision mixers receive instructions from the director, telling them when to cut from one picture source to another, producing a smooth sequence of images. The vision mixer sits at a console into which all the picture sources are fed. This console can produce effects (such as 'dissolving' and 'wiping') to make the transition from one scene to another either more interesting or unobtrusive. The job calls for quick reactions and an excellent sense of timing. Trainees are usually recruited from among those already working in television. They need normal colour vision, good hearing and a good general education.

Visual effects designer

Many of those concerned with the visual side of television – set designers, costume designers, graphic designers, make-up artists – are creating a visual effect or illusion. The services of specialist visual effects designers are called upon when a special effect is needed, such as a burning skyscraper or a rocket launch. They may work with actors or stunt performers on a life-size set on which they have started a carefully controlled fire, for example, or they may use scale models. Visual effects designers need a good working knowledge of sculpture, model-making, painting, optics and pyrotechnics, together with an understanding of the principles of physics, chemistry and electricity. Many other visual effects are generated electronically. There are few permanent posts with television companies for visual effects designers. Most work freelance or for facility houses which provide services for other film and television companies. However, vacancies in this field are rare.

Wardrobe work

Wardrobe stock-keepers issue, maintain and index costumes for television productions. Wardrobe operatives perform such tasks as assembling and labelling garments and packing them up for dispatch.

Weather forecaster (see Meteorologist)

Writer

Most of those who write for television and radio are freelances. There is seldom full-time work, but many authors find that writing can be an interesting and well-paid sideline. It is specialised work with its own techniques, so it is a good idea to learn these either by reading books or taking a course on the subject. It is possible, occasionally, to break into script writing when an unsolicited script is accepted, but most writers employ a literary agent who knows the market. Writers are not always asked to produce original work. Sometimes they are asked to contribute episodes for long-running serials or make adaptations of novels or short stories.

Scripts should always be typed or word-processed (double spaced) and accompanied by a stamped, self-addressed envelope. Most organisations return scripts they cannot use, but it is very unwise to send your only copy of a work. The BBC publishes *Writing for the BBC* which provides further information.

Case Study

Sue Limb collected her fair share of rejection slips when she started out as a writer; however, she found that one job almost certainly leads to another.

'My first published work was *Captain Oates: Soldier and Explorer* (with Patrick Cordingly, for Batsford), for which I collected several rejection slips, as I did for *Up the Garden Path*, a comic novel. Indeed, just before it was accepted, my agent rang me and said, "I think we'd better give up with this one and put it in a drawer."

Rejection was also a feature of my early contact with the BBC. I sent them two plays on spec and they were rejected. A few years later I had an idea for a comedy series, wrote the pilot and sent it in. They were interested enough to ask me to rewrite it – nine times. At this point, I withdrew, hurt.

About a year later I had a better idea, for a serial based on the home life of the Wordsworths, *The Wordsmiths of Gorsemere: An Everyday Story of Towering Genius*. This needed only one rewrite and has now run for two series. I have also written two radio serials of the Izzy books, *Up the Garden Path* and *Further Up the Garden Path* (or *Love's Labours*). I am very happy writing for radio because it requires the active participating imagination of the listener.

Once launched with the BBC, the rest is easy. They will ring up and ask me to do compilations and guest appearances, and the Light Entertainment Department is now very sympathetic to my ideas. I have done masses of work for Schools Radio and even won a Sony for *Big and Little*.'

4 Getting started

Job advertisements invariably demand previous experience, but, as the cliché goes, you can't get a job without experience, yet you can't get experience without a job. So, how do you go about getting that foot in the door?

Ten ways to get ahead of the crowd

1. Watch and Listen

An obvious – but important – one to start with: familiarise yourself with programmes. Sample as wide a range of television and radio output as you possibly can and analyse it critically. Ask yourself why a programme works or why it doesn't, and what you would do to improve it. Get to know which companies, writers, producers and directors are responsible for which programmes.

2. Prove your interest

Appropriate qualifications are important, but try to develop skills in related areas to demonstrate your enthusiasm to potential employers. For example, write for a local newspaper or your school or college magazine, or join an amateur theatre group.

Case Study

Hospital Radio has proved an excellent training ground for many of today's top producers and presenters. It gives volunteers the chance to learn how to handle professional equipment, put together balanced programmes, work as a team and build a relationship with an unseen audience.

Hospital radio stations are independent charities run by dedicated volunteers. There are currently about 390 such services broadcasting to over 90 per cent of the hospital population in the UK. They are funded by donations from the local communities they serve, and most station members take an active role in fund raising. Volunteers are also expected to do a degree of 'ward-walking' – collecting requests and meeting patients – as well as broadcasting.

There are also about 20 hospital television services in the country that broadcast either live (via cable) or by pre-recorded tapes to local hospitals. Most hospital services are members of The Hospital Broadcasting Association, which organises conferences, offers technical advice, and helps train volunteers.

As with professional broadcasting, hospital radio and television involve working long and unsocial hours, so you will have to make sacrifices if you want to become involved. It is not uncommon for new members to drop out once they realise they can't just come and go as they please!

3. Find out as much as possible

Conferences and events can be great places to make contacts and to find out more about the industry. If you know what you want to do, then talk to someone already doing that job.

Read as much as you can, not just about your own area of interest, but also on general questions on the current state of the broadcasting industry. Subscribe to a trade paper if you can afford it. If not, enquire about trade papers at your local library and read the media pages of the national press so you become familiar with the names of the major figures. Books on television and radio pour off the presses. Many are on the pricey side, so, again, it's well worth a visit to your local library. But beware: always check the year of publication. Many of the books which inhabit library shelves have been sitting there gathering dust for years and will not necessarily offer an accurate picture of the industry.

4. Visit the workplace

Get a flavour of the environment. Visit a studio, either on an organised tour, or simply by applying for audience tickets. Better still, write and ask for a period of work experience. This can provide a great opportunity to glimpse the industry from the inside, to make contacts and to demonstrate your skills.

Channel 5 provides some work experience placements, while Channel 4 offers neither work experience nor vacation employment. The BBC's policy varies from region to region. Most ITV companies operate a work experience programme, although they usually insist that applicants are students following a recognised course of study (such as a degree) at a college or university and that the student is either resident or attending a course in the transmission area of the company offering the attachment. Some companies will also ask applicants to undergo a written test or an interview before making an offer, but even if applicants are successful at this stage, a placement is still not guaranteed.

Case Study

David, who runs a small independent television company, believes that it is up to the individual to make the most of any work experience placement.

'Every year, I receive dozens of requests for work experience. But because we are a small company, it is very difficult to deal with these cold calling letters which just say "I want to work in television". It helps to have a target, to know what kind of job you want.

You have to show a willingness to fit into the team. The key qualities are probably personality, a sense of humour and a willingness to work twice as hard as the next person. We have had people in here on work experience who showed no interest or enthusiasm for what we were doing; I got the impression they were just looking for somewhere to hide for a week.

On the other hand, we have had some very good experiences. A couple of years ago, we were working on a documentary series and we brought in a guy from a training scheme as an extra pair of hands. He was very keen, he fitted in well, and at the end of his six-month placement we gave him a job.'

5. Keep your eyes and ears open

Once you do get on the inside – even if it is just for a few weeks – keep your eyes peeled. You can learn an enormous amount just by watching how jobs are done, learning who does what jobs, and the names of key people in the company. By simply turning up on time and doing your job (however menial) to the best of your abilities, you are already showing yourself to be dependable and enthusiastic. Make yourself indispensable. You may find yourself eligible to apply for the great many posts that are not advertised externally.

6. Develop transferable skills

Typing, word-processing and computer literacy are all skills worth acquiring. Even the fact that you have a full, clean driving licence will enhance your CV (see below). These abilities will also help enormously if you ever need to look for work outside the industry.

7. Make an effort with your CV

It is vital that your CV (curriculum vitae) makes a good impression, because television and radio stations receive hundreds of them every month; a few lines scrawled on some coffee-stained A4 will impress nobody.

Type your CV if you possibly can. It is worth paying to have this done professionally, as a good typist will know how to set it out and might be able to advise on content and phrasing. Your CV should be no longer than three A4 sides and should include:

- Full name and address.
- Date of birth.
- Schools attended.
- Examinations passed (dates and grades).
- Any other honours won at school or college.
- Any position of authority held at school or college.
- Training courses, colleges attended and qualifications

gained, together with dates and grades.
- ◆ Previous jobs held or any other experience (give names of companies and dates)
- ◆ Names and addresses of two referees whose permission you have previously obtained. One of these should be a previous employer or someone (such as a teacher) who has personal knowledge of your abilities.
- ◆ Personal interests and hobbies.
- ◆ Languages – mention if you can speak a foreign language and state level of fluency.

8. Market yourself

Let the employers know you exist.

Speculative letters

Don't be afraid of sending in a speculative letter – you have nothing to lose (apart from the price of a stamp, of course). However, do be aware that neither the BBC nor Channel 4 accepts speculative *job applications*. Make your letter look good and do your research. Find out all you can about the organisation you are applying to, or, if you are writing to a particular producer, make sure you are thoroughly familiar with their work. Nothing is guaranteed to irritate a producer more than a letter which (a) spells his or her name incorrectly and (b) clearly demonstrates that the writer has never watched or listened to any of their programmes in his or her life.

If you don't know the name of the person you should be writing to, phone the company and find out. In general, you should address your letter to the Chief Personnel Officer or the Head of Recruitment, but your letter is much more likely to receive attention if it is addressed to a specific individual. Enclosing a stamped, self-addressed envelope will also greatly enhance your chances of a reply.

State precisely where your abilities and interests lie: for example, if it is music, say what kind; if it is technical, state your qualifications and experience. Keep your letter short and enclose a copy of your CV.

Any letter must be neat, legible and interesting. So, at the risk of stating the obvious, here are some guidelines:

◆ Make a rough draft of your letter to be sure that it contains all the essential points.

◆ It does not matter if you cannot type your letter, but you must write neatly and legibly.

◆ Use good quality, preferably white, writing paper and a matching envelope.

◆ If you are writing in answer to an advertisement, mention where you saw it.

◆ If you write 'Dear Mr/Mrs Johnson', sign off 'Yours sincerely'. If you have been unable to track down a name, then start 'Dear Sir or Madam' and sign off with 'Yours faithfully'.

◆ Print your own name under your signature.

◆ Always show the finished letter to a teacher, tutor or reliable friend – especially if you have any doubts about spelling or grammar.

◆ Keep a copy of your letter for reference.

Demo tapes

Local radio stations, both BBC and IR, are always on the lookout for new talent of all kinds. All station managers receive a great many unsolicited demo tapes, so if you are thinking of sending one in, then you must keep it short: absolutely no more than ten minutes. An all-rounder's tape could include:

◆ News reading (two minutes, approximately 250 words).

◆ A well-structured interview (approximately three minutes).

◆ A sample of how *you* introduce music; remember, they already know what the music sounds like.

9. Take care with application forms

If you are replying to an advertised vacancy, take your time filling in any application form; a lazy or half-hearted effort could ruin your chances.

- ◆ Pay careful attention to any instructions on the application form; for example, if you are asked to complete the form in black ink, then *use* black ink (or you risk eliminating yourself before your application has even been read).
- ◆ Demonstrate how your skills match the *specific* qualities outlined in the job specification. Don't just rattle off your CV and hope for the best.
- ◆ Make several rough drafts before you fill in your application. Photocopy the original blank form and practise on that.

If and when you are called for an interview, it's only natural if you feel nervous or find it difficult to collect your thoughts. So it's always a good idea to think out beforehand what you would say in answer to these questions:

- ◆ Why have you decided to try for a career in radio/television?
- ◆ What made you apply for this particular job?
- ◆ What makes you think you will be good at this job?
- ◆ How would you like your career to develop? What would you like to be doing in five years' time?
- ◆ Why do you want to leave your present job (if you already have one)?

And, again, do your research – find out as much as you can about the company and the job itself.

10. Find out more

Dozens of organisations can offer further help and advice.

Skillset

Skillset is the Industry Training Organisation for the broadcast, film and video sectors. It operates at a strategic level within the industry, providing relevant labour market and training information, as well as encouraging greater investment in training. Skillset seeks to influence education and training policies to the industry's best advantage at both national and international levels, and also works to create greater and equal access to training opportunities and career development.

Radio Academy

This is the professional membership body for those working in, or with an interest in radio. Members receive a monthly newsletter and annual yearbook with a comprehensive radio industry directory. They are also entitled to reduced-rate admission to the various events organised by the Academy.

Networking

This is a membership organisation for women working or who are seeking work in any capacity in film, video and television.

BKSTS – The Moving Image Society

BKSTS is the technical society for film, television and associated industries such as sound. It organises a wide range of training courses, conferences, seminars and other events. It also accredits media courses within the further and higher education sector. The society publishes two journals: *Image Technology* and *Cinema Technology*, plus a monthly newsletter, *Images*.

BECTU's Student Link-Up Scheme

This scheme allows students to receive the *Stage, Screen and Radio* journal, and entitles them to ask for the union's assistance with projects. Full membership is available to graduates at a reduced rate.

WAVES (Women's Audio Visual Education Scheme)

A charity which provides film and video education, training and advice for women at all stages of their careers.

Addresses and phone numbers for all these groups can be found in Chapter 9.

5 Getting in and getting on

We asked some of the most respected figures in the television and radio industries to answer the following question: 'Based on your knowledge and experience, what is the most important piece of advice you would give to someone who is just starting out on their career?' Here is a selection of their replies.

John Birt, Director-General, BBC
'You need to be extremely enterprising when setting out on your career. Qualifications are important, but you also need to prove your interest and enthusiasm by finding out as much as you can about organisations you want to work for; contacting people you admire in your chosen field; asking for a week or two's work experience and meeting as many people as you can once you are there. Broadcasting is exciting and fulfilling – but there is no easy way in!'

Jill Dando, Broadcaster
'Although there are many television channels offering opportunities, don't expect overnight success. Be prepared to learn the trade in local television and radio stations. You can afford to make one or two mistakes there away from the national glare! Remember, the deeper the foundations, the stronger the structure.'

Jenny Abramsky, Director of Continuous News, BBC
'Take risks, be brave, work long hours, and above all, be creative and innovative.'

Trevor McDonald, Broadcaster
'The greatest asset you have is self-belief... the feeling that despite all the numerous difficulties you *can* succeed. And once you've achieved the first priority, which is actually getting a job, always, always strive to do better. Never give up striving and never give up.'

Sue Lawley, Broadcaster
'I think the most important technique to develop when setting out on a career in television or radio is not to be afraid to ask questions. This applies to whether you are in front or behind the microphone; if you don't understand the answer, then ask again. If you haven't understood, the chances are others haven't either. If you want to get on, there is no room for false pride.'

Nick Ross, Broadcaster
'Avoid clichés, forsake adjectives, be fair to views with which you disagree, and consider how *you* would feel if you were being portrayed, reported or edited in the manner you propose.'

Eamonn Holmes, Broadcaster
'Know what you want to do. I am not of the school of thought that believes "get into the business and you will find something that suits you." If you want to work in broadcasting you should have a pretty fair idea of the area you want to specialise in.

Knowing where you want to go shows you are more likely to be better informed and better motivated and hence progress more quickly. Then once you have decided on what your job is going to be, find out a little about everybody else's.

Failing all that, if in front of the camera is your chosen path, just make sure you've got a good agent!'

Michael Grade, Former Chief Executive, Channel 4
'1. Know what you want to do and be very specific.
2. Watch current TV and be able to converse critically and in detail about programmes.
3. Read the industry trade papers so you know who's who and what's what – and pick up the media vocabulary.
4. Don't give up.'

6 The future of the television and radio industry

It has never been easy to break into television or radio and, sadly, this situation is unlikely to change, as more and more people compete for fewer and fewer vacancies. However, other important developments should be expected over the next few years which will affect everyone working or seeking work in the industry.

The freelance market

The freelance market will continue to grow, as organisations slim down to a core number of key personnel, with extra staff recruited as work demands. However, a high percentage of freelances are employed for only short periods at a time and many work for less than half the year. This means that anyone looking for a job in television or radio will need to be mobile, financially astute and possess excellent self-marketing skills.

Digital broadcasting

Few people doubt the impact that digital broadcasting will have on the industry in the coming years. Essentially, this is a new way of condensing television and radio signals so that more, better quality pictures and/or sounds can be transmitted in the same space as one conventional analogue channel (although analogue transmissions are expected to continue alongside digital broadcasts for at least another ten to 15 years).

All broadcasters are gearing up for the new age of digital broadcasting. The BBC, for example, has planned a range of digital television services, including a 24-hour television news channel, which will be free to all viewers with suitable receivers. 'Side channels' will provide background information on programmes, while the output of the BBC's centres in Scotland, Wales and Northern Ireland will be available across the UK. Meanwhile, BBC Worldwide, the corporation's commercial arm, is developing a number of joint venture subscription channels.

Digital technology can also accommodate interactive services, such as home shopping and video-on-demand. British Interactive Broadcasting for example, a consortium of British Sky Broadcasting, BT, Midland Bank and Matsushita (Panasonic), plans to make hundreds of television channels and interactive services available to viewers with satellite dishes in 1998.

However, the launch of so many new services is unlikely to lead to a corresponding rise in job opportunities. Many of these channels will simply make use of existing archive material, while the high-volume but low-cost nature of original productions on cable and satellite means that any new jobs will not only be lower paid, but also multi-skilled.

Multi-skilling

In fact, the multi-skilling ethos has swept the entire broadcasting industry in recent times. The rapid development of user-friendly technology, such as desk-top editing and graphics packages, means that distinctions between roles have blurred significantly, and will continue to do so. This has dramatically altered the nature of many jobs and has led to cuts in all sectors of broadcasting. Everyone must now be prepared to adapt existing skills as well as learn new ones. Computer literacy will become an essential requirement for virtually all posts.

Media ownership

Another development with significant employment implications concerns media ownership– particularly within the ITV sector, which has seen a major concentration of power in the past few years. The three most important 'players' in the network are Carlton Communications, The Granada Group and United News and Media. At the time of writing, Carlton owns both Central and Westcountry, plus 20 per cent of both GMTV and ITN; The Granada Group owns LWT, and 20 per cent of both GMTV and ITN; while United owns Anglia, HTV, Meridian, 20 per cent of ITN – and 30 per cent of rival station Channel 5! Yorkshire Television, meanwhile, owns its smaller neighbour Tyne Tees (the company which broadcasts as Channel 3 North East). Further takeovers are inevitable. The overall effect of this activity has been a mass merging of resources with the loss of hundreds of jobs. However, many ex-employees have found work as freelances or with independent companies.

Multimedia

Finally, a word about multimedia. The ultimate impact of this technology is difficult to predict, although its potential is enormous. Many broadcasters, notably the BBC, are investing heavily in this area, and it is arguably one of the few sectors of the broadcast media where some form of employment growth is likely. Contact BIMA (The British Interactive Multimedia Association) for more information.

7 Qualifications available and choosing a course

There are hundreds of media-related courses and more appear every year, so it's essential that you choose one best suited to you and your interests. Don't assume that just because the word 'media' or 'communication' appears in the title that it will be exactly what you are after. Generally speaking, it is best to opt for a course that will give you a solid grounding in practical skills.

Broadcasting organisations, while acknowledging that many courses are of a good standard, do not recognise them officially. If you are thinking of taking a course and hope to work for a particular company when you complete it, you should check with the company's recruiting manager that the course has a good reputation and that the qualification would enhance your chances of being recruited.

National Vocational Qualifications (NVQs)

NVQs and their Scottish equivalents SVQs are practical qualifications designed to prove that a person can do a job, rather than just talk about it or pass an exam in that subject. There are five levels of NVQ/SVQ awards: Level 1 represents basic skills, whereas a Level 5 award designates a high standard of professional ability. For the broadcast, film and video sectors, qualifications for Levels 2, 3, 4 and 5 have been, or are currently being, developed.

NVQs and SVQs for these sectors are monitored, administered and awarded by Skillset, the Industry Training Organisation, working in association with the Open University Validation Service. These qualifications are becoming increasingly important in an industry which has traditionally lacked formal training structures.

Choosing a course: checklist

These guidelines have been compiled in association with Skillset and BKSTS: The Moving Image Society. It is a good idea to try and find out as much as you can *before* you apply, rather than, say, asking questions on the day of your interview.

♦ Read any course publicity very carefully. Try to visit the college or institution where the course is taking place and talk to current students and tutors. Find out what has happened to previous students and trainees.

♦ Find out if formal qualifications are more important than experience or enthusiasm.

♦ Always ask about course fees. These will vary according to the duration and type of the course, but some can be very expensive. Don't assume that a course is necessarily better just because it charges higher fees.

♦ Find out how much practical work (ie, hands-on experience) is involved. Ideally, this work should continue throughout the course.

♦ Is the equipment you will be using of broadcast standard? How old is it and how does it compare to that which you would use in a work environment? Will you have unrestricted access to this equipment, or will you have to share it with other courses? What level of competence are you likely to have achieved by the end of the course?

♦ How much technical support is available? Will there be a technician on hand to repair faulty equipment?

♦ Find out if it will be possible to gain NVQs/SVQs during your training, or if the course is based on NVQ/SVQ standards.

♦ How many students will be on the course and how much

personal tuition can you expect? How much time is supervised?

◆ Do the staff have personal experience within the industry? Do they keep up to date with current industry practices? Find out if industry representatives visit to give regular talks and seminars.

◆ Is a work experience placement part of the course and what is it likely to involve? Is it up to the college or the student to find the placement? Is it assessed?

◆ What will you have to produce for your overall course assessment, and how will it be assessed?

Where to study

Course listings

It is almost impossible to provide comprehensive, up-to-date information on the many and widely differing degree courses available. The best advice is to use this chapter as a jumping-off point and to consult *Directory of Further Education Courses* (published by CRAC), *Post Graduate Directory of Graduate Studies* (CRAC), *British Qualifications* (Kogan Page) and the *UCAS Handbook*. The address and phone number of the organisations mentioned here can be found in Chapter 9.

Media Courses UK, edited by Lavinia Orton, published by the British Film Institute (BFI), contains detailed information on courses available throughout the UK, including entry requirements, syllabus content and the qualifications they can lead to. The BFI also publishes *A Listing of Short Courses in Film, Television, Video and Radio*, an annually updated booklet covering short courses available throughout the UK.

The Radio Academy's web site includes a list of UK radio courses. The address is http://www.radacad.demon.co.uk/

Course databases

Skillset/BFI Database

The Skillset/BFI Database contains details of long and short courses throughout the UK, including undergraduate and postgraduate courses in further and higher education. This information can be accessed from a number of outlets across the country, including all the regional consortia (see below). Queries should be kept as specific as possible. However, an entry on the Skillset database should not be taken as an endorsement of the quality of any given course.

CRCA Database

The Commercial Radio Companies Association (CRCA) also maintains an extensive database of current training courses. Contact the CRCA directly for further information.

Broadcast journalism courses

If you are interested in a career in broadcast journalism, then contact the National Union of Journalists, or the Broadcast Journalism Training Council, enclosing a large sae, for details of approved courses and course requirements.

Consortia

Skillset, in association with other relevant bodies, has established a network of national and regional consortia across the UK. These are independent bodies which are in regular contact with the key trainers and educators in their respective areas. Their objectives include identifying training needs, coordinating and, where appropriate, initiating provision for, developing and implementing NVQs/SVQs.

In areas where consortia have yet to be established, Regional Arts Boards or Film Councils will have details of local workshops and courses.

Contact details for national and regional consortia can be found in Chapter 9.

Entry-level courses and training schemes

ARTTS International

ARTTS International (Advanced Residential Theatre and Television Skillcentre) provides intensive, residential, one-year, ten-week and two-year courses in acting, directing or production operations for television, theatre, video/film and radio.

CSV Media

CSV Media is the largest media training agency in the UK. It works in partnership with over 100 local TV and radio stations and offers nationally validated training in a range of media skills. CSV trains unemployed young people and adults at over 40 sites nationwide, concentrating on practical, on-the-job experience.

Contact CSV for details of your nearest course.

Film and television freelance training (ft2)

ft2's New Entrant Technical Training Programme is a full-time, two-year course which trains new entrants to become junior production and technical grade assistants in specific areas of the film and television industries. These include editing, sound, and hair and make-up, but ft2 does *not* train scriptwriters, directors or producers. The scheme is heavily over-subscribed, so the selection procedure is extremely tough. A further initiative, the Setcrafts Apprenticeship Training Scheme, was launched in 1996, in association with Skillset. This trains a small number of young people seeking to establish careers as freelance carpenters, fibrous plasterers and set painters in the industry.

For further information, contact ft2 enclosing an A4 stamped, self-addressed envelope.

First Film Foundation

First Film Foundation is a charity providing training and support to new writers, producers and directors based in the UK and

Ireland who are working on feature film and TV drama projects. Their current schemes include North by Northwest, a training programme for new screenwriters, and New Directions, an American showcase scheme for new directors. First Film also offers a script evaluation service which, for a fee, provides a written assessment of TV and film drama scripts submitted by new writers.

Gaelic Television Training Trust

The Trust offers a two-year programme of college and industry-based training to Gaelic speakers. This includes a year-long attachment, either in Glasgow with BBC Scotland or Scottish Television, or in Aberdeen with Grampian Television. Although not a formal requirement, it is expected that applicants will hold some further or higher education qualification. Selection is on the basis of interview, with competence in Gaelic taken into consideration when allocating places.

London International Film School

The London International Film School offers a practical, two-year diploma course to professional levels, accredited by the British film technicians' union, BECTU. Approximately half of each term is devoted to film production, and half to practical and theoretical tuition. Applicants should have a degree or an art or technical diploma. Other qualifications may be accepted in cases of special ability or experience. All applicants must submit samples of their work and be proficient in English. Courses begin in January, April and September.

National Film and Television School

The School offers full-time postgraduate and post-experience courses of three years' professional training (screen music and screenwriting two years), enabling its graduates to take positions of responsibility in all aspects of film and television production.

Courses are offered in animation direction, cinematography, documentary direction, editing, fiction direction, producing screen design, screen music and screenwriting. The average age of entry is 25, although there is no hard and fast rule on age. Candidates need to demonstrate a knowledge of basic skills in practice and theory within their specialist area.

WITCH (Women's Independent Cinema House)

WITCH runs video/media production courses with Open College accreditation, which are open exclusively to women aged between 16 and 25. Courses last for 15 weeks and no previous experience or formal qualifications are required.

Video Engineering and Training (VET)

VET's courses specialise in newer technologies. They are intended for people with some basic knowledge who wish to update their skills.

9 Useful addresses

BBC

BBC Radios 1, 2, 3, 4 and 5 Live, Broadcasting House, Portland Place, London W1A 1AA; 0171 580 4468

BBC Television Centre, Wood Lane, London W12 7RJ; 0181 743 8000

BBC World Service, Bush House, PO Box 76, Strand, London WC2B 4PH; 0171 240 3456

BBC Corporate Recruitment Services, Villiers House, The Broadway, Ealing, London W5 2PA; 0181 849 0849

BBC Television Training, BBC Elstree Centre, Clarendon Road, Borehamwood, Herts WD6 1JF; 0181 953 6100

BBC Monitoring, Caversham Park, Reading, Berkshire RG4 8TZ; 01734 472742

BBC/Open University Production Centre, Walton Hall, Milton Keynes MK7 6BH; 01908 274033

BBC National Regions

BBC Northern Ireland, Broadcasting House, Ormeau Avenue, Belfast BT2 8HQ; 01232 338000

BBC Scotland, Broadcasting House, Queen Margaret Drive, Glasgow G12 8DG; 0141 339 8844

BBC Wales, Broadcasting House, Llantrisant Road, Llandaff, Cardiff CF5 2YQ; 01222 572888

BBC English Regions

Broadcasting Centre, Pebble Mill Road, Edgbaston, **Birmingham** B5 7QQ; 0121 414 8888

Broadcasting House, Whiteladies Road, **Bristol** BS8 2LR; 0117 973 2211

Broadcasting Centre, Woodhouse Lane, **Leeds** LS2 9PX; 0113 244 1188

New Broadcasting House, PO Box 27, Oxford Road, **Manchester** M60 1SJ; 0161 200 2020

Broadcasting Centre, Barrack Road, **Newcastle upon Tyne** NE99 2NE; 0191 232 1313

York House, Mansfield Road, **Nottingham** NG1 3JB; 0115 955 0500

St Catherine's Close, All Saints Green, **Norwich**, Norfolk NR1 3ND; 01603 619331

Broadcasting House, Seymour Road, Mannamead, **Plymouth** PL3 5BD; 01752 229201

Broadcasting House, Havelock Road, **Southampton** SO14 7PU; 01703 226201

ITV

Anglia Television Ltd, Anglia House, Norwich NR1 3JG; 01603 615151

Border Television plc, Television Centre, Carlisle CA1 3NT; 01228 25101

Carlton Television, 101 St Martin's Lane, London WC2N 4AZ; 0171 240 4000

Central Broadcasting, Central Court, Gas Street, Birmingham B1 2JT; 0121 643 9898
Carlton Studios, Lenton Lane, Nottingham NG7 2NA; 0115 986 3322

Channel Television Ltd, The Television Centre, St Helier, Jersey JE1 3ZD, Channel Islands; 01534 68999

Channel 3 North East, The Television Centre, City Road, Newcastle upon Tyne NE1 2AL; 0191 261 0181

GMTV (Good Morning Television) Ltd, The London Television Centre, Upper Ground, London SE1 9LT; 0171 827 7000

Grampian Television plc, Queen's Cross, Aberdeen AB15 4XJ; 01224 846846

Granada Television Ltd, Quay Street, Manchester M60 9EA; 0161 832 7211

HTV, The Media Centre, Culverhouse Cross, Cardiff CF5 6XJ; 01222 590590.
The Television Centre, Bath Road, Bristol BS4 3HG; 0117 977 8366

ITN (Independent Television News), 200 Gray's Inn Road, London WC1X 8XZ; 0171 8333 3000

ITV Network Centre, 200 Gray's Inn Road, London WC1X 8HF; 0171 843 8000

LNN (London News Network), The London Television Centre, Upper Ground, London SE1 9LT; 0171 827 7700

LWT (London Weekend Television), The London Television Centre, Upper Ground, London SE1 9LT; 0171 620 1620

Meridian Broadcasting Ltd, Television Centre, Southampton SO14 0PZ; 01703 222555

Scottish Television plc, Cowcaddens, Glasgow G2 3PR; 0141 300 3000

Tyne Tees Television – as for Channel 3 North East

UTV (Ulster Television plc), Havelock House, Ormeau Road, Belfast BT7 1EB; 01232 328122

Westcountry Television Ltd, Western Wood Way, Langage Science Park, Plymouth PL7 5BG; 01752 333333

Yorkshire Television Ltd, The Television Centre, Leeds LS3 1JS; 0113 243 8283

Channel 4, S4C and Channel 5

Channel 4 Television, 124 Horseferry Road, London SW1P 2TX; 0171 396 4444

S4C, Parc Ty Glas, Llanishen, Cardiff CF4 5DU; 01222 747444

Channel 5, 22 Long Acre, London WC2E 9LY; 0171 550 5555

Non-BBC national radio stations

Atlantic 252, 74 Newman Street, London W1P 3LA; 0171 637 5252

Classic FM, PO Box 3434, London NW1 7DQ; 0171 284 3000

Talk Radio, 76 Oxford Street, London W1N 0TR; 0171 636 1089

Virgin Radio, 1 Golden Square, London W1R 4DJ; 0171 434 1215

National and regional consortia

Midlands: Midlands Media Training Consortium, Studio 11, Nottingham Fashion Centre, Huntingdon Street, Nottingham NG11 3LF; 0115 993 0151

North East England: Mediaskill Development Ltd, PO Box 2NE, Newcastle Upon Tyne NE99 2NE; 0191 232 5484

Northern Ireland: Northern Ireland Film Commission, 21 Ormeau Avenue, Belfast BT2 8HD; 01232 232444

North West England: North West Media Training Consortium, Campus Manor, Childwall, Abbey Road, Liverpool L16 0JP; 0151 722 9122

Scotland: Scottish Screen Training, 4 Park Gardens, Glasgow G3 7YE; 0141 332 2201

South West England: Skillnet South West, 23 Trenchard Street, Bristol BS1 5AN; 0117 925 4011

Wales: Cyfle, Llawr Uchaf, Gronant, Penrallt Isaf, Caernarfon, Gwynnedd LL55 1NW; 01286 671000

Other useful addresses

ARTTS International, Highfield Grange, Bubwith, North Yorkshire YO8 7DP; 01757 288088

BECTU (Broadcasting Entertainment Cinematograph & Theatre Union), 111 Wardour Street, London W1V 4AY; 0171 437 8506

BFBS – *see* SSVC (Services Sound and Vision Corporation)

BFI (British Film Institute), 21 Stephen Street, London W1P 2LN; 0171 255 1444

BIMA (The British Interactive Multimedia Association), 5–6 Clipstone Street, London W1P 7EB; 0171 436 8250

BKSTS: The Moving Image Society, 63–71 Victoria House, Vernon Place, London WC1B 4DA; 0171 242 8400

British Sky Broadcasting, 6 Centaur's Business Park, Grant Way, Isleworth, Middlesex TW7 5QD; 0171 705 3000

Broadcasting Standards Commission, 7 The Sanctuary, London SW1P 3JS; 0171 233 0544

Broadcast Journalism Training Council, The Secretary, 188 Lichfield Court, Sheen Road, Richmond, Surrey TW9 1BB; 0181 940 0694

Cable Communications Association, Fifth Floor, Artillery House, Artillery Row, London SW1P 1RT; 0171 222 2900

Commercial Radio Companies Association (CRCA), 77 Shaftesbury Avenue, London W1V 7AD; 0171 306 2603

Community Radio Association, The Work Station, 15 Paternoster Row, Sheffield S1 2BX; 0114 279 5219

CSV Media, 237 Pentonville Road, London N1 9NJ; 0171 278 6601

Film and Television Freelance Training (ft2), Fourth Floor, Warwick House, 9 Warwick Street, London W1R 5RA; 0171 734 5141

First Film Foundation, 9 Bourlet Close, London W1P 7PJ; 0171 580 2111

Gaelic Television Training Trust, Sabhal Mor Ostaig, Sleat, Isle of Skye IV44 8RQ; 01471 844373

Hospital Broadcasting Association, Staithe House, Russel Street, Falkirk FK2 7HP; 01324 611996

Intelfax Ltd, 140–142 Lower Marsh, London SE1 7AE; 0171 928 3044

Intermedia, 19 Heathcoat Street, Nottingham, NG1 3AF; 0115 950 5434

IRTC (Independent Radio and Television Commission – Ireland), Marine House, Clanwilliam Place, Dublin 2; 01676 0966

ITC (Independent Television Commission), 33 Foley Street, London W1P 7LB; 0171 255 3000

London International Film School, 24 Shelton Street, Covent Garden, London WC2H 9HP; 0171 836 9642

MTV, Hawley Crescent, London NW1 8TT; 0171 284 7777

National Film and Television School, Beaconsfield Studios, Station Road, Beaconsfield, Buckinghamshire HP9 1LG; 01494 671234

National Union of Journalists, Acorn House, 314–320 Gray's Inn Road, London WC1X 8DP; 0171-278 7916

Networking, Vera Productions, 30–38 Dock Street, Leeds LS10 1JF; 0113 242 8646

PACT (Producers' Alliance for Cinema and Television), 45 Mortimer Street, London W1N 7TD; 0171 331 6000

QVC, Marco Polo House, 346 Queenstown Road, London SW8 4NQ; 0171 705 5600

Radio Academy, PO Box 4SZ, London W1A 4SZ; 0171 255 2010

Radio Authority, Holbrook House, 14 Great Queen Street, Holborn, London WC2B 5DG; 0171 430 2724

Radio Ireland, Radio Ireland House, 124 Upper Abbey Street, Dublin 1; 01804 9000

Raidió na Gaeltachta, Baile Na nGall, Traighli, Co Chiarrai; 066 55114

Reuters Television, 85 Fleet Street, London EC4P 4AJ; 0171 250 1122

Royal Television Society, Holborn Hall, 100 Gray's Inn Road, London WC1X 8AL; 0171 430 1000

RTE (Radio Telefís Éireann), Donnybrook, Dublin 4; 01208 3111

Skillset, 124 Horseferry Road, London SW1P 2TX; 0171 306 8585/8457/8143

Sky Television – *see* British Sky Broadcasting

SSVC (The Services Sound and Vision Corporation), Narcot Lane, Chalfont Grove, Gerrards Cross, Bucks SL9 8TN; 01494 874461

Teletext Ltd, 101 Farm Lane, London SW6 1QJ; 0171 386 5000

TnaG (Teilífís na Gaeilge), Bail na hAbhann, Conamara, Co na Gaillimhe; 091 593636

Video Engineering and Training Ltd (VET), Northburg House, 10 Northburg Street, London EC1V 0AH; 0171 490 4001

WAVES (Women's Audio Visual Education Schemes), 4 Wild Court, London WC2B 4AU; 0171 430 1076

WITCH (Women's Independent Cinema House), Blackburn House Centre for Women, Hope Street, Liverpool L1 9JB; 0151 707 0539

10 Further reading

Magazines and journals

It is vital that you stay up to date with who's who and what's what, so always keep an eye on the media pages of the national press (for example, *The Guardian* and *The Independent* on Mondays), or better still, if you can afford it, subscribe to a trade paper. Some of the more accessible publications include:

Audio Visual (monthly; multimedia communications)
Broadcast (the weekly trade paper of the television and radio industry)
Campaign (weekly; advertising)
The Stage and Television Today (weekly; targeted at actors)
TV World (ten issues a year; international sales and distributions)
UK Press Gazette (weekly; journalism and the media)

General information

For general information about working in the media, try:

The Skillset Careers Handbook, produced by the Industry Training Organisation for the broadcast, film and video sectors. Available directly from Skillset (see Useful

Addresses, page 77) on receipt of an A4 stamped, self-addressed envelope.

The Official ITV Careers Handbook, published by Headway/ Hodder & Stoughton. Order from bookshops, or directly from the ITV Network Centre.

Lights, Camera, Action! by Josephine Langham; published by British Film Institute (BFI).

The Guardian Media Guide.

Production companies

For details of UK production companies, consult:

The Broadcast Production Guide (International Thompson)
Kemp's International Film and Television Yearbook (Reed Information Services)
The Knowledge (PA Publishing Co. Ltd)
PACT Members Directory (PACT)
BFI Film and Television Handbook (BFI) contains recent industry data, plus a directory of thousands of cinema and television contacts. All are updated regularly and are available from many public libraries.

Recommended reading

Birtwistle, Sue and Conklin, Susie, *The Making of Pride and Prejudice*, Penguin/BBC (1995)

Blum, Richard A, *Television and Screen Writing: From Concept to Contract* (3rd edn), Focal Press (1995)

Boyce, Ed, Crisp, Mike and Jarvis, Peter, *Editing Film and Videotape*, BBC

Chantler, Paul and Harris, Sim, *Local Radio Journalism* (2nd edn), Focal Press (1996)

Chater, Kathy, *The Television Researcher's Guide*, BBC

Goodwin, Peter, *Television and the Tories: Broadcasting Policy Under Thatcher and Major*, BFI (1996)

Hayward, Anthony, *The Making of Moll Flanders*, Headway (1996)

Holland, Patricia, *The Television Handbook*, Routledge (1997)

Home, Anna, *Into the Box of Delights*, BBC (1993)

Jarvis, Peter, *The Essential Television Handbook*, Focal Press (1996)

Keith, Michael C, *The Radio Station* (4th edn), Focal Press (1997)

Keyes, Jessica (ed.), *The Multimedia Handbook*, McGraw-Hill

Mansfield, John, *News! News!,* BBC

McLeish, Robert, *Radio Production* (3rd edn), Focal Press (1994)

Millerson, Gerald, *Effective TV Production* (3rd edn), Focal Press (1993)

The Multimedia Yearbook, Macmillan

Newby, Julian, *Inside Broadcasting*, Routledge (1997)

Pines, Jim (ed.), *Black and White in Colour*, BFI (1992)

Prior, Allan, *Script to Screen: From Z Cars to The Charmer*, BFI (1996)

Rowlands, Avril, *The Television PA's Handbook* (2nd edn), Focal Press (1993)

Seddon, Peter, *Where's the Designer*, BBC

Thompson, Chris, *Non-Linear Editing,* BFI/Skillset (1994)

The Videomaker Handbook, Focal Press (1996)

Watkinson, John, *An Introduction to Digital Audio*, Focal Press (1994)

Yorke, Ivor, *Television News* (3rd edn), Focal Press (1995)

Index

The Kogan Page *Careers in...* series

Careers in Architecture *(5th edition)*

Careers in Computing and Information
   Technology *(new title)*

Careers in Journalism *(8th edition)*

Careers in the Police Service *(5th edition)*

Careers in Retailing *(6th edition)*

Careers in Teaching *(7th edition)*

Careers in Television and Radio *(7th edition)*

Careers in the Travel Industry *(6th edition)*

Careers Using English *(new title)*

Careers Working with Animals *(8th edition)*